Focus on Data Center

My eyes established a vision for my life,

After seeing the beauty of Lord Shri Krishna (with belief, devotion and love).

My senses started working intelligently,

After understanding the morals of Lord Shri Krishna's stories and his past times (with belief, devotion, and love).

My hands started writing good books,

After folding hands in front of Lord Shri Krishna (with belief, devotion, and love).

Oh Lord Krishna, what am I without your blessing,

Oh Lord Krishna, your blessing is the reason for my breath, for my actions, for health, for prosperity, for peace and everything that I possess.

Hare Krishna Hare Krishna, Krishna Krishna Hare Hare,

Hare Rama Hare Rama, Rama Rama Hare Hare,

Table of contents

Register This Book and Get a Gift!

Thank you very much for purchasing this book.

First of all, before you read ahead, please **register your book** here:

https://servicemanagers.org/dc-registration

As a **ServiceManagers.Org DC Book Club** member, you will receive lots of benefits and support from our Book Club:

https://servicemanagers.org/dc-book-club

Certification Program

This is the official book for taking the ServiceManagers.Org Certified **Data Center Professional**™ (**DCP**™) program:

https://servicemanagers.org/data-center-certification

This Career Program will improve your job opportunities and certify your skills gained from this book.

The Certification Exam is online, and once you sign up you can take it whenever you want (no deadline limits). Just sign up today for your exam, and start enjoying this book.

Preface

Data centers plays an integral role in IT organizations; they either make or break the IT organizations. Despite data centers being like the heart and brain of IT organizations, there are very few books on data centers. If there are some, they are very technical and difficult to be understood by beginners and management professionals of IT.

A few years back when I had some meetings with some IT managers, experienced professionals, and consultants, it was a shock for me that they had no idea on basic data center environment, terminology, and operations. When I had surveyed their interest on this area, they were very inquisitive to learn and also saddened that they couldn't find any easily understandable materials on this subject area.

Hence, here is a book on data centers giving an overview on data center operations and my consulting knowledge in data center management which I have been writing for the last three years.

The focus is on topics like basics of data center terminology, checklists for data center operations, roles and responsibilities in data center management, essentials for data center management operations, metrics, data center setup, data center migration, data center site selection, auditing data center, due diligence questions for data center management, green data center and best practices, and 50 plus interview questions for data center management jobs.

This book will be very useful for the beginners like people who have just joined in IT, experienced professionals who have very basic knowledge on data centers, and management people who can get the most important epitome on data center operations and environment.

In my opinion, the prerequisite for reading this book is that the reader should have basic understanding in information technology and should be a working professional in IT.

Finally, if you are planning to gain comprehensive knowledge on data center management or aspiring to find a job in data center management, this book is for you.

Acknowledgments

If I have to start thanking people name by name,

I should first thank my mother Krishna Kumari Pabbathi, for giving birth to me, with all the organs and body in good condition.

I should thank my father Surya Prakash Pabbathi, for teaching me the right morals, values, and placing me in one of the best school, sacrificing his comforts and pleasures.

I should thank my grandfather Satyanarayana Pabbathi, who iterated that good books are the greatest friends, and exemplified his life for principles.

I should thank Balasaraswathi Penmetsa, for teaching Mathematics patiently in my 10th class, when I was dumb at Mathematics, and when no professional teacher could teach me with patience.

I should thank my graduation friend Venkatesham Padigela, who inspired me to become a topper in my graduation.

I should thank my first spiritual alma mater - ISKCON and my first spiritual mentor Kalakanta Prabhu, for making me understand life and Lord Krishna (which is the best thing that ever happened in all my life). Also, I would like to thank my other mentors Pravardhana Acharya Prabhu and Madan Sundar Prabhu who always show their mercy by giving great spiritual knowledge to me.

I know that I am still missing numerous people who have taught, helped, and supported me numerous times in my life.

Hence, I would give my highest respects and surrender to the Supreme Lord Krishna for arranging all these wonderful people in my life, and giving me so many lessons, experiences, and knowledge!!!

Big thanks to **Angel Berniz**, and **ServiceManagers.Org** for publishing another book.

About the Author

Kiran Kumar Pabbathi has worked for various companies in the IT industry which gave him detailed insight of ITAM, ITSM, ITIL best practices and EAM.

Kiran has had the privilege to work in different roles taking care of service desk operations, request fulfillment, incident management, availability management, capacity management, information security management, IT service continuity management, sharepoint administration, project management, ITIL consulting and trainings, and ITAM Consulting.

Kiran is a certified professional in ITIL® Expert, PRINCE2® (Foundation and Practitioner), Six Sigma Green Belt, ISO/IEC20K – Foundation, Cloud Computing - Foundation, TMAP – Foundation (Test Management Professional), MCP in SharePoint 2003 Customizations, MCTS in MS Project 2007 and trained in CCNA & MCSE. Kiran is also an accredited ITIL trainer and Invigilator for BCS, UK.

His other authoring works include:

- "PDCA for ITIL – Metrics, CSFs and workflows for implementing ITIL practices" published by TSO, UK (ISBN 9780117082076) which gives a direction for implementing ITIL processes and designing ITSM solutions.
- "Charm of friendship" published by Pothi, India (ISBN 9789382715924) explaining the importance of friendship for children.
- "Guidance for ITAM – Step by step implementation guide with workflows, metrics, best practices and checklists" published by ServiceManagers.Org (ISBN 9780991320509)

- "Guidance for EAM – Processes, implementation steps, workflows, best practices, and checklists" published by ServiceManagers.Org (ISBN 9780991320514)
- "Focus on IAM - CSFs, metrics, checklists, best practices, and guidelines for defining IAM processes and implementing IAM solutions" published by ServiceManagers.Org (ISBN 9780991320530)
- "Guidance for Incident Management - According to ISO/IEC 20000 & 9001 Standards, Six Sigma and ITSM Best Practices" published by ServiceManagers.Org (ISBN 9780991320547)
- "Guidance for Problem Management - According to ISO/IEC 20000 & 9001 Standards, Six Sigma and ITSM Best Practices" published by ServiceManagers.Org (ISBN 9780991320554)
- "Guidance for Change Management - According to ISO/IEC 20000 & 9001 Standards, Six Sigma and ITSM Best Practices" published by ServiceManagers.Org (ISBN 9780991320561)

Introduction to Data Center

Basic Terminology

Application Server: Server designed to install, operate, and host applications and associated services. They are designed to power frameworks which are involved in extensive processes related to a specific set of software

Automatic Transfer Switch: (ATS) detects when utility power is outside of an acceptable range (to activate the UPS and generators) and when the utility power resumes and switches from UPS to utility power.

Availability Objective: Ability of a component or service to perform its required function at a stated instant or over a stated period of time.

Bit: A unit of measurement of information.

Byte: A sequence of 8 bits (enough to represent one character of alphanumeric data) processed as a single unit of information

Best Practice: Practices that are well recognized and which have proved the ability for demonstrating success in respective areas (e.g. PMBOK for project management, ITIL for IT service management, MOF for IT service management, COBIT for IT governance, etc.).

Blade Server: Blade server is a stripped down computer server with a modular design optimized to minimize the use of physical space, housing multiple thin, modular electronic circuit boards, known as server blades. Each blade is a server in its own right, often dedicated to a single application.

Bridge: A device that connects two local-area networks (LANs), or two segments of the same LAN, providing different network services.

Cabling System: Structured cabling within the Data center where the copper and fiber cabling is used as the typical media and are terminated via several types of connectors. Common components include fiber housings, patch panels, multimedia boxes, and data faceplates.

Carbon Footprint: It is the amount of CO2 emissions that are produced in Data center operations. Data centers consume 25% or more of the electric bill.

Change Implementation Plan: Plan that details the complete coverage on change implementation explaining who should own the change, who will implement the change, who will test the change, what are the prerequisites for implementing change, how the change has to be implemented, definition of back out plan, etc.

Communication Plan: Plan which defines how the communication has to be disseminated to the organization stakeholders with respect to business messages and individual messages.

Command Center: A small area to be designed within the Data Center with a desk, chair, raised glass wall (to lower the noise pollution) and a landline phone.

Content Delivery Networks: An arrangement of servers to deliver content at top speed for users in different geographic locations and reduce the time of content delivery for the user.

Cooling System: Refers to the chillers and air handlers used to regulate ambient temperature and control humidity within the Data center. This system might incorporate the air conditioning system used to cool regular office space within the same building, known as house air, or might be independent of it.

Computer Room Air Conditioner: CRAC/CRAH (pronounced crack) that uses a compressor to mechanically cool air/Computer Room Air Handler (pronounced crah) that uses chilled water to cool air.

Critical Success Factor (CSF): Vital principles necessary for the success of a business; it is based on CSF's and metrics are developed.

Close Coupled Cooling: In fact, it's a type of data-center air-conditioning system that brings the cooling source as close as possible to the high-density computing systems that generate the most heat. Instead of cooling down the entire room, close-coupled cooling systems located in a rack cool the hot air generated by the servers on just that rack.

Chiller: it's a machine that uses chilled water to cool and dehumidify air in a data center. Of all the components of a data center's air conditioning system, this is the one that consumes the most amount of electricity -- as much as 33% of a data center's power.

Colocation: A colocation (colo) is a data center facility in which a business can rent space for servers and other computing hardware.

Compliance: Compliance is either a state of being in accordance with established guidelines, specifications, or legislation or the process.

Conduit Space: The vertical or horizontal space in a building to provide access to connections between computers or operation centers, power, chilled water, fiber, telephone, junction box, and other infrastructure support.

Connected Load: The loads for the actual operating values of the equipment connected.

Data Center: A facility that centralizes computer systems, servers, and networking systems to store, process, manage, and disseminate huge volumes of information through the use of redundant data communications, backup power supplies, environmental controls, and security devices.

Data Backup: Backup is the activity of copying files or databases so that they will be preserved in case of equipment failure or other catastrophe. Backup is usually a routine part of the operation of large businesses with mainframes as well as the administrators of smaller business computers.

Data Center Bridging: An architecture that is based on a collection of open standards Ethernet extensions developed through the IEEE 802.1 to improve and expand Ethernet networking and management capabilities in the data center.

Data Center Design: Design which models the Data center's IT resources and architecture layout.

DCIE: It is the ratio of the IT equipment energy to the total data center energy use.

Dedicated Servers: In the web hosting business, a dedicated server refers to the rental and exclusive use of a computer that includes a web server, related software, and connection to the Internet, housed in the web hosting company's premises.

Distributed Data Center: The data centers that are geographically distributed, automated, orchestrated and interconnected with high speed/ultra-low latency network.

Equipment Distribution Area: The space where access layer equipment such as LAN/SAN/KVM switches and servers are located.

Fire Suppression: Devices associated with detecting or extinguishing a fire in the Data Center. The most obvious components are water-based sprinklers, gaseous fire suppression systems, and hand-held fire extinguishers.

Gateways: An internetworking device that can convert the format of data in one computing environment to a format that is usable in another computer environment, and which acts as the backbone of any network architecture.

Gigabyte: the gigabyte is a multiple of the unit byte for digital information storage. The prefix giga means 1,000 megabytes or 1,000,000,000 (1 billion) bytes in the International System of Units

Ghost Server: Server that is underutilized or not being used in a Data center.

Gigabit Interface Converter: Transceiver that converts electric currents (digital highs and lows) to optical signals, and optical signals to digital electric currents.

Green Data Center: Data center where the mechanical, lighting, electrical and computer systems are designed for minimum energy consumption and minimum environmental impact.

Grid Computing: Computing that involves large number of computers in a grid to collectively accomplish large tasks. It is an extension of cluster computing which reduces costs and offers business advantage over other traditional approaches.

Hardened Shell: Terms associated with the ready status of a building such as 18 - 22' floor height, exterior wall resistance to blast/storm/wind resistance, roofing and roof drainage, and uniform live load.

High Performance Computing Cluster: Computers that are usually physically very close together & connected together with a network and centrally coordinated by some special software.

Horizontal Distribution Area: The space where aggregation layer equipment such as LAN/SAN/KVM switches are located.

Hot Aisle/Cold Aisle: A layout design for server racks and other computing equipment in such a way that it eliminates the mixing of hot air (from servers) and cold air (from air conditioning) which is one of the biggest contributors to inefficiencies in the data center.

In-room Electrical: In-room electrical refers to all power-related facilities within the Data center. This normally includes electrical panels, conduits, and several types of receptacles.

Inter Process Communication (IPC): It is a mechanism by which two processes exchange information and coordinate activities on a network.

Impact: Measure that defines the number of users affected by an outage. Usually denoted as 1,2,3,4.

Key Performance Indicator (KPI): Vital metrics necessary for an organization to meet its business goals, which reflect the CSFs of an organization.

Kilobyte: unit of information equal to 1000 bytes.

Line Conditioner: Device that keeps the power supply at a constant voltage and frequency.

Main Distribution Area: The space where core layer such as routers, LAN/SAN switches, and PBX's are located.

MAID: Storage technology that uses a large group of hard disk drives, hundreds or even thousands, in which only those disk drives in active use are spinning at any given time. MAID reduces power consumption and prolongs the lives of the drives.

Metric: Measurements that quantitatively evaluate the performance of IT operations.

Megabyte: Unit of information equal to 1000 kilobytes 1,000,000 (1 million) bytes.

Mobile Rack Servers: Rack mounted servers that are built on cabinets which allows them to be upgraded more flexibly.

Network Interface Cards (NIC): NIC's are printed circuit boards that are installed on workstations which provide the physical connection between the network cable and the workstation.

Network Operating Center: NOC - place from which administrators supervise, monitor and maintain a telecommunications network.

Power Distribution Unit (PDU): This electrical device is used to control the distribution of power to the individual loads. Control may be as simple as a series of switch or circuit breakers to interlocked logic operating solid state relays.

Physical Space: Physical space refers to the footprint that Data Center-related items occupy. This generally applies to the overall area of the Data Center and its associated spaces, such as electrical rooms or storage areas.

Policy: Policy is a management directive that significantly influences processes and procedures.

Priority: Measure used to identify the criticality of an incident/problem/change. Usually denoted as 1, 2, 3, and 4. Calculated through the formula, Priority= (Impact + Urgency)/2

Process: A set or sequence of activities that results in achieving an output or business goal.

Power Usage Effectiveness (PUE): It is the ratio of the total data center energy use to total IT energy use. PUE is the inverse of the DCIE metric.

Rack: A data center rack is a type of physical steel and electronic framework that helps in housing the servers, managing space efficiently, providing cable management, and HVAC. Typically, approximately 16 sq ft for the rack and 24 sq ft of support area for a total of 40 sq ft.

Rack server: Is a computer dedicated to use as a server and designed to be installed in a framework called a rack.

Raised Flooring: Raised flooring is an elevated grid system that is frequently installed in large Data Centers. Cooled air, electrical whips, and data cabling are routed through the space under the raised floor, promoting better air flow and enabling easier management of power and cable runs. Water pipes, fire suppressant cylinders, moisture detectors, and smoke detectors can also be located here as well.

REC: Renewable energy certificates or renewable energy credits. RECs are tradable commodities that show that 1 megawatt-hour of electricity was purchased from a renewable source, such as solar, wind, biomass or geothermal. An increasing number of companies are buying RECs to offset the amount of electricity generated from fossil fuels that their data centers consume.

Redundancy: The duplication of critical components of a system with the intention of increasing reliability of the system, usually in the case of a backup or failsafe. N+1 refers to the initial component need plus 1 backup, N+2 refers to the initial component need plus 2 backups, etc.

Recovery Point Objective: The maximum amount of active transaction data loss, expressed in units of time.

Recovery Time Objective: It is the amount of time that can elapse after a business interruption until the service needs to be restored.

Redundancy: Replication is the process of sharing information so as to ensure consistency between redundant resources, such as software or hardware components, to improve reliability, fault-tolerance, or accessibility.

Repeaters: Repeaters were created to regenerate and amplify weak signals, thus extending the length of the network. The basic function of a repeater is to re-amplify the data signal to its original level.

Resources: Any tangible assets (like IT infrastructure, people, money, or anything) that will be helpful to deliver an IT service.

Router: A highly intelligent network device that is primarily used for large networks and provides the best data path for effective communication.

Server Platform: It is the software in a server, acting as an engine driving the server. It is usually used in conjunction with an operating system.

Serial Advanced Technology Attachment: Serial ATA is a serial link, with a single cable that has four wires creating a point-to-point connection between devices.

Service: Provision of value to customers without any ownership of risks and internal costs associated in developing a service.

Skinless Servers: Servers that are designed with a reduced amount of rack, chassis and, in some cases, even motherboard components to maximize server density potential and reduce material use and power consumption.

Standby Power: Standby power includes all backup power systems responsible for support of the Data Center's electrical load in the event that normal utility power fails for any reason. This system traditionally includes large batteries, known as an uninterruptible power source or uninterruptible power supply, and one or more generators.

Standard: Rules and conventions that help to implement policies and enforce required conventions.

Storage Area Network: An architecture to attach remote computer data storage devices to servers so the devices appear as locally attached to the operating system.

Surge Protector: A device designed to protect against jumps in voltage from the power source.

Switch: Network Device that routes an input signal to an output signal – such as PCs to each other, PCs to servers, PCs to printers, etc.

Terabyte: A terabyte is a multiple of the unit byte for digital information storage and is equal to 1012 (1 trillion short scale) bytes or 1000 gigabytes.

Transceiver: A device that both transmits and receives signals along the network cable and connects a computer to the network.

Tier level: Tier Level is a standardized methodology used to determine availability in a facility.

Uninterrupted Power Source (UPS): Uninterruptible Power Supply, or battery backup, is an electrical apparatus that provides emergency power to equipment usually when utility mains fail.

Unit: A measurement which determines the amount of equipment that can be housed in rack - one "u" is 1.75 "in height."

P2V server management: Physical to Virtual (P2V) server management is defined as products that manage the hypervisor and all virtual machine instances that run on servers.

Virtual Machine Hypervisor: Hypervisor which enables the operation of multiple operating systems (OS) instances concurrently on a single physical server.

Virtual Private Server: Server on which multiple domains are hosted (on the same server) running its own operating systems that can be independently booted.

Virtual Server: Server that is portioned into sub-machine (through the help of virtual configuration) providing the users with root access and full computing power.

Virtualization: A stripped down server computer which runs multiple operating systems on a single machine with a modular design optimized to minimize the use of physical space and energy.

Virtualized Data Center: Data center with some or all the hardware virtualized.

White Box Server: A data center computer that is not manufactured by a well-known brand name vendor.

Data Centers in Organizations

Data center, in simple words, means a place where all the data is centered, and where all the data is stored, accumulated and disseminated. It is a facility that centralizes computer systems, servers, storage devices and networking systems to store, process, manage, and disseminate huge volumes of information through the use of redundant data communications, backup power supplies, environmental controls and security devices. Data center is also called as a server farm, a centralized repository for the storage, management, and dissemination of data and information.

Data centers are physical or virtual infrastructure used by enterprises to house computer, server and networking systems and components for the company's information technology (IT) needs, which typically involve storing, processing and serving large amounts of mission-critical data to clients in a client/server architecture.

Data centers manage the computer systems which manage the most critical services and intellectual property like hosting organizations websites, managing mailbox servers, management of financial records, etc.

Data centers can be operated as: In-house Data centers, Colocation Data centers, and Managed Data centers.

In-house Data center is a Data center that has been developed for business purposes in the same organization. Here all infrastructure, people, environment, etc. is maintained by the same organization and its human resources.

Colocation (Colo) is a data center facility in which a business can rent space for servers and other computing hardware. Generally, a Colo provides the building, cooling, power, bandwidth and physical security while the customer comes in with servers and storage equipment.

- Can be an entire building dedicated to a single tenant where a Colocation company provides managed facility and other services.
- Can be a demised suite within a building where the tenant occupies the space exclusively (shared or dedicated infrastructure) with tenants in other suites within the building.
- Can be a fenced off cage within a larger space occupied by other tenants or a place where one or more lockable racks are placed in a multi-tenant space.

Advantages of opting colocation Data center:

- It can lower the cost of ownership, risk and time associated with new construction
- It avoids large capital expenditures and moves funding to operational model
- It achieves a data center reliability that may be difficult to maintain in a client-built and managed facility
- It provides an opportunity to improve IT operational processes, controls, and documentation
- It will make the company carefree from real estate driven imperatives

- It provides an opportunity to run the data center function to a deep-skilled, capable provider

Managed services Data center is similar to colocation, only the provider owns the server, while performing defined business tasks in the operation and maintenance of the equipment. Managed services include including offsite data backups, firewall & VPN management, virus scanning, intrusion prevention, integrity monitoring, etc.

Advantages of opting managed services Data center are:

- Access to staff with specialist skills.
- Greater control of service levels and performance.
- Provides a single "converged" connection, resulting in cost-savings on infrastructure.
- It is an extremely flexible - a pay-as-you-go payment plan that allows for quick growth when necessary, or cost savings when you need to consolidate.

Data centers can be classified into three types: Standard Data Center, Build to Suit Data center, Custom Design Data Center.

- Standard Data Center

  A data center that generally involves a design that follows common industry standards and best practices. It usually covers the layout of the rows of cabinets, typically capable of supporting a moderate power density.

- Build-to-Suit Data Center

The "Build to Suit" or "Turn Key" or "Move-In Ready" are used by some data center builders and providers in the industry. It generally offers a somewhat lower level of customization within certain limits of a basic standard design.

- Custom Design Data Center

  In custom design data center, some extra features are provided in the form of modularity, flexibility, reliability, energy efficiency and power density, or any of the combinations.

Data center types are classified into 4 tiers as: Tier 1, Tier 2, Tier 3, and Tier 4. The tier classifications describe the infrastructure topology required to run the Data center operations.

Tier 1 Data center: Is a basic Data center that has non-redundant components, composed of single path for power and cooling distribution providing 99.67% availability (28.8 hours of interruption/ year). Any unplanned outage, failure of any system, or human errors can cause disruption to Data center operations.

Tier 2 Data center: Is a Data center that has partial redundancy and a single, non-redundant distribution path. It can provide availability up to 99.75% (22 hours of interruption/ year). Any unplanned outage, failure of any system, or human errors may impact the Data center operations.

Tier 3 Data center: Data center that has redundant capacity components (N+1) and multiple independent distribution paths serving the computer equipment. All IT equipment's are dual powered and also has 12 hours of on-site fuel storage for N capacity. With this configuration, it is possible to manage maintenance periods without affecting the continuity of service on the servers. It can provide availability up to 99.98% (1.6 hours of interruption/ year)

Tier 4 Data center: It is a fault tolerant Data center that has fully redundancy (2N+1) in terms of electrical circuits, cooling, and network. This architecture can withstand even the most serious of technical incidents without server availability ever being affected. It can provide availability up to 99.99% (0.8 hours of interruption/ year).

Data center tier classification summary

Parameters	**Tier 1**	**Tier 2**	**Tier 3**	**Tier 4**
Availability %	99.67	99.75	99.98	99.99
Active capacity components	N	N+1	N+1	N after any failure
Distribution paths	1	1	1 Active & 1 Alternative	2 Simultaneously active
Concurrently maintainable	No	No	Yes	Yes

Fault Tolerance	No	No	No	Yes
Cooling	Load density depended	Load density dependent	Load density dependent	Class A

Equipment in Data center

Some of the important equipment that make up the data center infrastructure can be mentioned as servers, racks, routers, switches, modems, firewalls, cables, hard drives, tape drives, backup storage resources, power generators, cooling towers, Uninterrupted power supply system (UPS), application delivery controllers, WAN optimizers, security scanners, wireless access points, shredder, SAN storage, tape libraries, network attached storage, etc.

Racks

Rack is a DC equipment which organizes the DC assets in an efficient manner considering space, air, power, etc. Racks improve power protection, cooling, cable management, device management, physical security, mobility, and ease of installation. Types of racks can be classified as: Open frame racks, Rack enclosures, Wall mount racks.

Open Frame Racks: Racks that are with open frames with mounting rails, but without sides or doors. They are good for high-density cabling because they provide convenient access and lots of open space for cable management.

Rack Enclosures: Racks that have removable front and rear doors, removable side panels, and four adjustable vertical mounting rails. They're good for heavier equipment, hotter equipment and higher wattages per rack.

Wall Mount Racks: Racks that can be attached to the wall, saving floor space.

Servers

Servers can be classified into three types as blade server, tower server, and rack server.

Blade Server

An architecture that places multiple servers on the blades in a single chassis. It is widely used in Data centers to save space and improve system management. The blade server combines the hardware of the computer server, such as processors, memory cards and network connections, in one extension card (also known as one server blade). A blade server in this sense is literally a server on a board containing processors, memory, integrated network controllers, an optional fiber channel host bus adaptor (HBA), and other input/output (IO) ports. Server blades typically come with one to four processors and integrated Ethernet. Blade servers typically have a front panel containing a number of informational LEDs, relating to power and system activity

Tower Server

A tower server is similar in size and shape to the cabinet for a tower-style personal computer consisting processors, memory card slots, network connections, hard disk drives, etc.

Rack Server

Is a computer dedicated to use as a server and designed to be installed in a framework called a rack (also called a rack mounted server). A rack server is in contrast to a tower server, which is built into an upright, standalone cabinet. Even this server consists of processors, memory card slots, network connections, hard disk drives, etc.

A single rack can contain multiple servers stacked one above the other, helping to manage the space efficiently, providing better cable management, and providing better HVAC. The rack server configuration also simplifies cabling among network components.

Cables

Cabling in DC can be classified into two types as passive cabling and active cabling.

Passive cabling are always copper-based cables that provides direct electrical connection between corresponding cable ends.

Active cabling can use either copper wire or fiber optics to provide the links through optics and/or electronics within the connectors. Furthermore, these fiber media are classified into single mode and multimode.

Application Delivery Controllers

Application delivery controllers enable customers' applications to be highly available, accelerated and secure; they are an evolution of traditional server load balancers (SLBs) which provide high availability, performance, and security.

They are network devices that helps sites direct user traffic to remove excess load from two or more servers and are typically located between the firewall/router and the web server farm.

Functionality:

- uses algorithms to determine how traffic is distributed
- inspect packet headers for keywords or file types, and direct the request to the appropriate server based on this information.
- provide real-time and historical analysis of all user and network traffic
- employs an array of mechanisms to improve application performance.
- serve as the gateway to the network and authenticate each user attempting to access an application.

WAN Optimizers

WAN optimizers are networking devices that prioritize traffic, and guarantee a certain amount of available bandwidth for mission critical applications; they reduce congestion, improve the speeds of file transfers, and make applications more reliable.

Functionality:

- prioritizes bandwidth to ensure that mission critical applications across the WAN gets the fastest performance.
- accelerates and encrypts WAN traffic between data centers.

Router

An internetworking device that enables communication between two or more logical networks. It is derived from the word "Route", which is meant for finding the best route for networks. Functionalities of a router can be mentioned as internetwork communication, best path selection, packet switching and forwarding.

Routers can be classified into two types: Hardware router and Software router.

Hardware routers act as an interface between different LAN's. They can be classified into two types - fixed and modular router. In fixed routers you cannot add/modify interfaces. In a modular router you can add/modify interfaces.

Software routers act as an interface inside the LAN.

Switch

Switches create a virtual circuit between two connected devices, establishing a dedicated communication path between two devices. They network to provide micro segmentation which allows maximum utilization of the available bandwidth. A switch is also able to facilitate multiple, simultaneous virtual circuit connections.

Fabric switch is a device which connects multiple network devices. These switches have 16 to 32 ports, which can connect 16 to 32 machine nodes.

Bridge

A bridge is a Layer 2 device used to divide, or segment, a network. It is capable of collecting and selectively passing data frames between two network segments.

Bridges do this by learning the MAC address of all devices on each connected segment. Using this information, the bridge builds a bridging table and forwards or blocks traffic based on that table.

This results in smaller collision domains and greater network efficiency.

Repeater

Repeaters are the networking devices that are created to regenerate and amplify weak signals, thus extending the length of the network. The basic function of a repeater is to re-amplify the data signal to its original level. These days they are built-in to hubs or switches.

Flywheel UPS

Flywheel is form of mechanical (kinetic) energy storage, which is analogous to the potter's wheel concept. This energy is stored by the movement of a rotor to spin on its own axis.

UPS systems that use the flywheel energy storage are called Flywheel UPS. Flywheel UPS are reasonably costlier but have longer life, simple maintenance, and a considerably less carbon foot print.

Biometric Security Devices

Devices that verifies the person's physical characteristics and allows the person to access a specific location or infrastructure device which includes finger print scanners, face recognition, retinal scanners, palm scanning, etc. These technologies can be implemented as:

Cabinet authentication units: Authenticates a user to specify which cabinet they are attempting to access using biometric, keypad, and card methods.

Server cabinet locks: Cabinet locks provide advanced access controls with the help of locks and biometrics.

Door locks: Door locks authenticate the person through the help of biometric, keypad and card methods at the entrance of Data centers and secure kiosks in Data center.

Virtual Tape Libraries

A virtual tape library (VTL) is a storage system consisting of hard disk drives (HDDs) that acts like a physical tape library. VTLs are typically used for backup, recovery, and data archiving purposes

A virtual tape library solution can perform backups and recoveries at a much faster speed. Speeding up the backups and recoveries will greatly improve the efficiency of Data center operations.

SAN

Storage Area Network connects multiple servers and storage devices on a single network using Fiber Channel connections and block protocols (SCSI).

SANs allow data centers to pool storage, eliminating the isolated storage arrays attached to servers. It reduces the excess capacity required for each server enabling administrators to centralize storage management.

NAS

Network attached storage is a shared storage on a Local Area Network which uses standard network and file sharing protocols.

NAS devices are stand-alone, having their own OS, networking, integrated hardware, and software. NAS devices offer shared storage, accessible from a number of platforms. NAS devices can easily be added on to an existing network and each storage device has its own network interface.

Data Center Asset Lifecycle for Beginners

Data center equipment lifecycle starts with an equipment request placed by the user/team who needs the equipment. Once the equipment requisition is approved, data center operations team would consult the user/team to discuss the specifications of the equipment and the available data center room space.

After the discussions with operations team, the user/team submits a data center space request form. Accordingly, the data center manager, operations manager, and facilities manager would discuss how and where to accommodate the new equipment, its environmental issues (i.e. equipment BTU and power specifications), space, and floor location. Once the data center manager approves the request, the equipment requisition is sent to the procurement team.

Once the procured equipment is received, the inventory/facilities team receives and stocks the equipment. When large and heavy equipment is involved, special arrangements has to be done for receiving the order.

Once the equipment is received the respective facilities/operations manager will be notified of the equipment arrival.

Then comes the installation of equipment in the data center with respect to cabinet design, cabinet/rack, UPS, KVM, etc.

Installation of the equipment happens as per the guidelines of data center architect and operations manager. Exceptions can happen only when it is specified by the server manufacturer dictating the equipment must be housed in their proprietary cabinet.

After the installation is done, operations will label the cabinet (both front and rear) with the unique floor grid location and with the power circuit serving that unit. Equipment spacing within the cabinet/rack should allow appropriate airflow for proper cooling and blanking panels will be installed to fill in all vacant rack space.

The operations team then identifies the hardware equipment with the appropriate fully qualified server names, and will label all equipment within the cabinets so that hardware is easily identifiable.

Removal of old hardware is coordinated with the Data center manager, Facilities Manager and change management process, following all policies, standards, and guidelines relating to data destruction, wiring removal, and component disposition. When a new system is replacing a pre-existing machine, the old system must be properly decommissioned via the Change Management process.

Roles and Responsibilities in Data Center Management

Project Manager

Responsibility addresses oversight of designated people, processes and projects that involve multiple processes within the organization. Project managers will usually have responsibility for outcomes across the multiple processes as assigned. He/She manages the activities of an assigned IT process including designated projects or programs, which includes planning, coordinating, administering, and evaluating programs, projects, processes, procedures, systems, standards, and/or service offerings.

Responsibilities of Project Manager

- Hire, orient, train, evaluate and develop employees. When necessary carry out disciplinary actions.
- Assign personnel to various tasks, direct activities, and monitor progress to ensure objectives are achieved.
- Demonstrate effective team management and supervisory skills including planning, priority setting, conflict resolution, and managing multiple efforts concurrently.
- Provides oversight and guidance to staff for a variety of projects within the Data Center. Supervises staff including prioritizing and assigning work; conducts performance evaluations; ensures staff is trained; ensures that employees follow policies and procedures; maintains a safe working environment; and, makes hiring, termination, and disciplinary recommendations.
- Develops and monitors budgets.

- Serves as a liaison with internal departments, executives, external organizations and public representatives. Represents the organization at a variety of meetings, public events, on committees, and/or other related events regarding services, programs, projects, and facilities.
- Prepares, reviews, interprets, and analyzes a variety of information, data, and reports. Recommends and implements tactical procedures to improve operational effectiveness and enhancements to services or programs.
- Tracks assigned budget and resources and provides input to the annual budget development process. Prepares cost estimates and budget recommendations. Monitors expenditures.
- Assists in planning and the development of short term goals. Coordinates operational and program efforts in alignment with goals and objectives related to assigned area of responsibility.
- Approves risk management plans and the gathered requirements
- Manages the budget to meet or exceed goals and comply with applicable accounting requirements.
- Ensures that the project plan is reviewed and approved
- Reviews the progress of project
- Communicates with the customer team
- Manages human resources with respect to extension of contracts, release, etc.
- Reviews metrics analysis

Note: Responsibilities can differ based on organization's requirements.

Required Qualifications and Skill Set

Skills

- Working experience and good knowledge of change management methodologies like:
    1. Leavitt's system model
    2. Burke-Litwin change model
    3. Kotter's 8 step change model

- Thorough understanding and knowledge on project planning and risk management for IT systems.
- Exceptional communication skills both written and verbal.
- Able to work effectively at all levels in an organization.
- Excellent active listening skills.
- Problem solving and root cause identification skills.
- Strong analytic and decision making abilities.
- Must be a team player and able to work with and through others.
- Ability to influence others and move toward a common vision or goal.

Must have:

- A bachelor's degree or master's degree in business administration or computer science or information technology.
- Thorough knowledge on project management methodologies like PMP or PRINCE2.

- Few years of experience in change management or transition management.
- Advanced Proficiency with Microsoft Word, Excel, Power Point.
- Excellent communication, documentation and presentation skills.
- Ability to negotiate, plan and, communicate change activities across the organization.
- Good analytical and systematic approach to problem solving.
- Excellent judgment, tact, and decision-making ability.

Note: Required skills can differ based on organization's requirements.

Operations Lead

As the leader of an operations team, the Operations lead will have a broad range of duties – including 24/7 on-call response and support, vendor oversight, interaction with development staff and consultants. The Operations lead will be responsible for ensuring continuous operations within a green-field, multi-building data center campus.

Responsibilities of Operations lead

- Prepare plans for development and installation of data center according to present and future requirements.
- Monitor all software and hardware products and ensure compliance to Data Center standards and systems.
- Administer installation and maintenance of organization data center.
- Monitor all activities related to analysis and implement all data center plans.
- Supervise designing and evaluation of all server systems according to measurement of process.
- Perform troubleshooting on data center, identify issues, and assist in timely resolution of the same.

- Manage and perform research to design new advanced data center.
- Ensure adherence to all established guidelines and management practices for employees.
- Coordinate with various departments and perform integration of critical systems into infrastructure.
- Coordinate with IT, enhancement, and business teams, and develop strategies to ensure achievement of data center capacity.
- Maintain expert knowledge on all applications and its use on systems for various critical systems.
- Ensure client satisfaction and provide optimal level of customer services for various departments.
- Maintain effective professional relationships with IT relationship managers.
- Maintain expert knowledge on all business procedures associate with assets and capacity.
- Monitor all functional resources required for growth and ensure appropriate working.
- Equipment maintenance tasks such as managing and installing custom upgrades for customers; replacing cables, components and accessories; maintaining records and drawing schematics of the center to detail what equipment is installed.
- Relay communication to the ODC stakeholders.

- Escorting customers through the center on tours, responding to customer inquiries, and coordinating with other organizations.
- Administrative duties include scheduling employee shifts, planning for capacity changes, maintaining corporate databases, and ensuring compliance with government regulations.
- Reviewing the requirements and other documentation.
- Identify the project for potential risks.
- Record the incidents reported by the ODC.
- Plan and schedule the testing activities.
- Prepare test plan documents.
- Plan metrics and KPI.

Note: Responsibilities can differ based on organization's requirements.

One word responsibilities

Facility Maintenance Program, Strategy/ Schedule / Management, Process / Procedures, Documentation / Logging / Reporting, Interior / Exterior Maintenance and Repair, Lighting / Plumbing / Electrical / Construction, Janitorial Services, Interior / Exterior maintenance, Trash / Recycling, Building Security, Security Officers, CCTV / Card Access / Biometrics / Portals / Locks / Keys, Elevators, Ground Water Station / Sewer Lift Station, Roof / Lightning

Protection System, Fire Alarm System and Extinguishers, Support Systems, Service Requests / Work Orders, Tenant Improvements, Project / Contractor Management, Communications Systems, Phone System / Cell Phones / Pagers, Servers / Workstations / Printers / Fax / Copiers / Plotters, Network Cabling / Circuit management, Fiber / Copper Conduit / Ladder Racking, Move, Add , Change activities, SOP Documentation, Document Library / Repository, Life / Safety, First Aid / Evacuation, Permits and Licenses, Inventory Management, Equipment / Spares, Shipping / Receiving / Loading Dock, Change Management, Operational Budget, Service / Support Contracts, Training programs

Required Qualifications and Skill Set

- Bachelor's degree or equivalent experience in Computer Science or Information Technology
- Minimum of five years management experience in critical environment facility operations and maintenance
- Proven skill and commitment to building and maintaining relationships with employees, customers, and vendors
- Proficiency with MS Office (Outlook, Word, Excel, Power Point) Project, and drafting programs such as CAD or Visio
- Demonstrated technical aptitude, especially related to critical environment facility systems (electrical, mechanical, controls, etc.)

- Developed problem-solving and analytical abilities
- Excellent project management ability: planning, estimating, scheduling, risk and resource management, analytical ability, management skills, mature and professional demeanor, strong critical thinking
- Excellent written and verbal communication skills
- Self-motivated and able to motivate others

Note: Required skills can differ based on organization's requirements.

Facilities Manager

A facilities manager is responsible for the management of facilities, he/she ensures that an organization has the most suitable working environment for its employees and their activities.

Facilities managers generally focus on using best business practice to improve efficiency, by reducing operating costs while increasing productivity. Facilities managers are involved in both strategic planning and day-to-day operations, particularly in relation to buildings and premises. Likely areas of responsibility include:

- procurement and contract management
- building and grounds maintenance
- cleaning
- catering and vending
- health and safety
- security
- utilities and communications infrastructure
- space management

Responsibilities of Facilities Manager

- Coordinate and manage relationships with vendors and support track.
- investigate availability and suitability of options for new premises.
- calculate and compare costs for required goods or services to achieve maximum value for money.
- plan for future development in line with strategic business objectives.

- manage and lead change to ensure minimum disruption to core activities.
- direct, coordinate and plan essential central services such as reception, security, maintenance, mail, archiving, cleaning, catering, waste disposal, and recycling.
- ensure that the building meets health and safety requirements and that facilities comply with legislation.
- Translate EHS/EMS/Security Systems and Green Initiatives to manage at the site level.
- plan best allocation and utilization of space and resources for new buildings, or re-organizing current premises.
- use performance management techniques to monitor and demonstrate achievement of agreed service levels and to lead on improvement.
- Respond appropriately to emergencies or urgent issues as they arise and deal with the consequences.

Skills Required

- Basic Knowledge of Unix/Windows/Backup will help for first level support in few projects. Also, supervising and coordinating work of contractors is needed.

Essentials for Data Center Management Operations

Essentials for Executing DC Operations

Essentials for execution of DC operations can be defined by the below activities:

1. Definition of governance structure, roles, and responsibilities
2. Definition of escalation procedures
3. Definition of DC operations procedures
4. Definition of DC relationship procedures
5. Definition of knowledge management procedures
6. Definition of auditing procedures

Definition of Governance Structure, Roles, and Responsibilities

Definition of governance structure, roles and, responsibilities ensures that DC achieves its objectives and goals clearly with clear segregation of who is supposed to do what.

Definition of Escalation Procedures

Definition of escalation procedures will ensure and will enable those concerned to get the issues noticed by the management at the right time and by the right people.

Escalation issues and can be defined as per the below mentioned table:

Issue	Level
Operational issues	Level 0 – Operations manager
IT availability issues	Level 0 – Operations manager
Process related issues	Level 1 – Process Manager
Human resource availability issues	Level 2 – Project Manager
Risk related issues	Level 2 – Project Manager
Project plan related issues	Level 2 – Project Manager

Definition of DC Operations Procedures

Definition of operations procedures can be classified into the below areas:

- Contract management
- Incident management
- Problem management
- Change and release management
- Configuration management
- License management
- Business continuity management
- Information security management
- Event management
- Facilities management
- Quality management
- Service level management and Reporting

Definition of DC Relationship Procedures

Relationship procedures should ensure primarily focus on:

- Regular interaction with the customer and stakeholders, by participating in meetings with customer with respect to project review, analysis, forecasting, and discussions to get better understanding of the needs.
- Understanding customer's technical, business environment, and their needs.
- Conducting daily/weekly/monthly performance reviews.

Definition of Knowledge Management Procedures

Knowledge management procedures should focus on:

- Documenting and profiling the work done and organizing lessons learnt sessions
- Establishment of a database for knowledge articles
- Organizing the knowledge articles with knowledge article registration number, keywords, error, version, created by, etc.

Definition of Auditing Procedures

Auditing procedures should focus on:

- Personnel procedures and responsibilities, address employee termination, cross-functional, and systems training

- Program change controls that are adequate to ensure that changes are tested and approved before being moved into production status
- Backup procedures that are adequate to minimize business interruption and protect against loss of data in the event of a disaster,
- Physical security controls that are adequate to prevent unauthorized access to computer center areas
- Environmental controls that are adequate to minimize hardware/software losses from fire or flood.

Metrics for Data Center Operations

Metrics for Data center operations are essential ingredients that assess and evaluate the performance of operational activities and will stand as a basis for improvement of Data center operations.

Metrics for Data center management operations can be described as below:

- Available rack space and used rack space
- Available storage space and used storage space
- Rack floor utilization
- Number of planned outages
- Number of hours in planned downtime
- Number of catastrophic failures
- Number of unplanned outages/incidents
- Number of hours in unplanned downtime
- Number of security incidents
- Number of power outages
- Total number of power outage hours
- Average time taken for procuring a server
- Average time taken for installing a server
- Deployment and Provisioning Speed: Time it takes to deliver new data center capacity, and the time it takes to turn up new servers and other IT gear, and to assign them with the appropriate data center resources.
- DCIE: Ratio of the IT equipment energy to the total data center energy use. DCIE shows the power used by a data center's IT equipment as a percentage of the total power going into the data center. A DCIE of 50% means that 50% of the total power used by a data

center goes to the IT equipment, and the other 50% goes to power and cooling overhead.

- PUE: Ratio of the total data center energy use to total IT energy use.
- HVAC System Effectiveness: Ratio of the IT equipment energy to the HVAC system energy.
- Supply Air Temperature and Return Air Temperature: Difference between the supply and return air temperature.
- Supply Air Humidity and Return Air Humidity: Difference between the supply and return air humidity.
- Airflow Efficiency: Overall measure of how efficiently air is moved through the data center, from the supply to the return, and takes into account low pressure drop design as well as fan system efficiency.
- Cooling System Efficiency: Overall efficiency of the cooling system (including chillers, pumps, and cooling towers) in terms of energy input per unit of cooling output.
- UPS Load Factor: Ratio of the peak load of the uninterruptible power supply (UPS) to the design value of its capacity.
- UPS System Efficiency: Ratio of the UPS output power to the UPS input power.

Best Practices for DC Operations

Best practices are those real practices that have delivered efficient, effective, and excellent results in the real DC operations.

Best practices for DC operations can be mentioned as below:

- **Regular Meetings with All The Stakeholders**
  Regular meetings with all the stakeholders to update the trends and patterns of issues in DC operations.

- **Maintenance of IT Owners and Technicians Contacts List**
  Maintenance of IT owners, technicians and service owner's contacts details, and being up to date is a must which will enable the change management staff to do the planning & coordination with other service transition stakeholders.

- **Selection of New ICT and Telecom Equipment**
  Selection of new ICT and telecom equipment considering energy efficiency performance (eco-rating, service level, energy star), temperature & humidity supported, compliance with green regulations (REACH and WEEE), power density and airflow direction.

- **Maintain Spare Parts**
  Always have spare servers, stock of sorted cables, (prepared) cabinet/rack space.

- **Data Should Have Multiple Levels of Protection**
  All the data in DC should at least have three copies, in different locations, and one should be in remote region for

disaster recovery procedures. At the same time keep at least one data copy offline.

- **Data Backup, Recovery and Archival Should Be Done as Per The Defined SLA's**
  Timely backup, restoration, and archival of data should be done very carefully as per the defined SLA's. Also special efforts should be taken to decide which storage technology is needed for critical services, physical data storage, security, RTO, RPO, etc.

- **Always Have More Floor Space**
  Effective usage of floor space in the Data center is very important aspect which can be executed by upgrading the existing servers to thinner height servers or sleeveless servers, will help to effectively use the floor space in terms of watts per square feet, servers per square feet, and even square feet per rack.
  Also initial investment in floor space at the planning time of Data centers can pay huge dividends in the mid-to-long term.

- **Accurate Tagging and Identification of Assets**
  1. All equipment and servers must be labeled with identification information. This identification should be legible and consistent for all data center and mission critical facilities (Label and tag the servers, cables, network ports, racks, storage containers, patch panels, etc. with unique identifiers). Also label anything that is broken or that has passed the EOL (e.g. flaky hard drives or server).

2. All rack and power strip equipment must have labels affixed that identify the circuit and power distribution unit to which each is attached.

- **Maintenance of Cables**
  1. Cut and throw away damaged or broken cables.
  2. Use Velcro, not cable ties.
  3. Keep cables long enough, but not too long.
  4. Use short (18" or 3') power cables when possible.

- **Regular Checks on The Complete DC Infrastructure**
  Frequent health checks and testing of all DC infrastructures (servers, networking devices, power and cooling equipment, cables, etc.) should be done regularly.

- **Any Changes to The Data Center Environment or Equipment Should Be Done after Taking Approvals**
  1. No floor tiles may be removed without the permission of the Data Center Management.
  2. Authorized electricians must perform all modifications to electrical service. All such changes must be authorized and coordinated by the DC manager. All electrical testing must be conducted during off-hours.
  3. No work may be done below the raised floor area without the approval of the DCFM manager or data center planner.

  4. No hardware, furniture, shelving or other materials may be removed or added to the data centers or

other mission critical facilities without the written request and approval of the DC manager.

5. Removal of all decommissioned equipment must be coordinated with the DC manager. The data center planner will schedule a removal date and coordinate the equipment's disposition with the owner.

- **Security and Safety Procedures Adherence**
    1. All computer cabinets and doors must remain closed at all times, even while they are being serviced.
    2. Industrial cleaning liquids or fluids may not be left unattended in the data center. No highly ammoniated or chlorinated products are allowed in the data centers.
    3. All decommissioned/abandoned equipment will be removed from the data centers within one day of notification or the process required by individual departments.

CSFs

- **Selection of Right Human Resources**
  Selection of right human resources with good experience in Data center operations, technical knowledge, analytical skills, communication skills, and customer service attitude are a must for Data center management staff; it can be ranked as a number one issue among all the CSFs in DC management. Qualified support for NOC, and maintenance staff for the mission critical systems like UPS, power distribution, back-up generators and cooling systems are key to ensuring continuous availability.

- **Clarity in Roles and Responsibilities**
  Establishing clear roles and responsibilities for all the staff is very vital for efficient operations. By defining, documenting, and assigning roles and responsibilities within your governance structure, you establish clear expectations for the individuals assigned to each role.
  The RACI chart is useful in clarifying roles and responsibilities.

- **Best Designs, Equipment Systems and Robust Architecture**
  Best designs and equipment systems should be implemented with good redundancy.

- **Data Center Site Selection**
  Data center site is one of the most important factors which is dependent on skilled labor, costs, availability, climate, taxes, electric supply, risks of natural disasters, etc.

- **Adequate Monitoring and Management**
  Monitoring Data center performance with respect to server performance (both physical & virtual), application performance, bandwidth, airflow, voltage, power, doors, video surveillance and environmental conditions is very important.

  Monitoring tools will enable the DC service provider to pick up the alarms in real-time, drill down to the exact issue and start troubleshooting it before the end user feels the impact.

- **Regular Training Sessions**
  Training sessions on policies, and procedures is a must which should happen at regular intervals.

  Most of the delays and discrepancies in DC management happen due to unawareness on process, policies, and procedures; hence, it is a mandatory objective for IT management to conduct training sessions which can bring thorough awareness to all stakeholders.

- **Defined Processes for ITSM and ITAM Operations**
  There should defined ITSM and ITAM processes as per the available IT best practice frameworks.

  And all DC operational activities should be based on these defined processes.

DC Assets Attributes

DC assets attribute is very important to document and track information about the assets effectively. It is very common that the DC staff generally thinks that documentation and maintenance of asset attributes is a time taking and tiring job, but it definitely saves a lot of time during maintenance and operations of assets.

Hardware asset attributes

CI Id: Unique serial number for this configuration item.

CI Name: Unique Name for identifying the configuration item.

CI Description: Description identifying the CI.

Status - (Active/Inactive/Decommissioned).

Device Type (Storage/Networking/Cables).

Manufacturer: Manufacturer of the equipment.

Model Name: Model of the equipment.

Model Number: Model number of the equipment.

Serial Number: An identification number showing the position of a printed or manufactured item in a series.

Asset Tag Id: Asset tag details of the configuration item.

IP Address: A unique string of numbers separated by full stops that identifies each computer using the internet protocol to communicate over a network.

Purchase Date: Date when the asset has been purchased

Purchase Invoice Number: It is the unique number that the supplier sends for assets purchased.

Warranty Start Date: Start date of the warranty.

Warranty End Date: End date of the warranty.

Depreciation Start Date: Start date for depreciation.

Depreciation End Date: End date for depreciation.

Retirement Date: Date when the asset should be retired.

Net Value: Value of the asset in its respective currency.

Depreciated Value/Write-off Cost:

Cost of replacement: Cost occurred for replacement.

Criticality: Rating that defines criticality of the asset.

Number of Users: Number of users associated/dependent on the asset.

Relationships: Number of relationships defined to other assets.

Applicable Service Level: Service levels agreed as per the defined SLA's.

Availability: Defined availability as per the SLA.

Redundancy: Defined redundancy level.

Contingency: Defined continuity level.

Recovery Timeframe Required: Defined recovery timeframe.

Backup Device ID & Location: ID of the backup device & its location.

Software Asset Attributes

Product Name: It will be the software name

Base Version: Last version released by the manufacturer

Current Version: Current version released by the manufacturer

Patch Version: Version of the patch installed

Installation Date: Date when the installation was done

Installed By: Mentions who installed it or how it was installed (by which server/client machine)

Update Date: Date when the update was done

Updated By: Mentions who updated or how it was updated (by which server/client machine)

Criticality (L/M/H): Criticality of the software – Low, Medium, High

Related CIs (Documents, Application, Hardware, Network etc.)

License Type: Mentions the type of license (User based license, Floating license, Node locked license, etc.)

License Number: Number of the license

Number of Licenses #

Licenses Used #

License Validity: Validity of the license

Link to License Document: Mentions the path to access license document

Service Provider Name:

Service Provider contact details:

Server Name

IP Address

Link to DSL

Database Detail

Link to Source Code

Information Sensitivity - H / M / L

Purchase Cost / Date

Invoice Number

PO Number / Date

Upgrade Cost / Date

Finance Record ID

Network Assets Attributes

CI Id: Unique serial number for this configuration item

CI Name: Name of the networking device

CI Description:

Status - (Active/Inactive/Decommissioned): The current status of the equipment

Manufacturer: Manufacturer of the equipment

Model Name: Model of the equipment.

Model Number: Model number of the equipment.

Serial Number

Asset Tag ID

Private IP Address

Public IP address

Purchase Date

Purchase Invoice Number

Warranty Start Date

Warranty End Date

Depreciation Start Date

Depreciation End Date

Retirement Date

Connectivity Type

Network ID

LAN ID

WAN ID

Port Number

Address Type

Prefix Length

Subnet Mask

Alias Address

Group Addresses

Circuit Number

Access Speed

Shaped Speed

Destination Tel Number

Management IP

IP Address

IGMP

Database

Number of Slots

Net Value

Depreciated Value/Write-off Cost

Cost of replacement

Redundancy

Contingency

Disaster Recovery Plan Link

Last Disaster Recovery Test Date

Backup Device ID & Location

Useful Templates for DC Environment

Health Check Template

Health check has to be performed by the DC operator's every day to capture a comprehensive picture of the DC assets and its health status.

Here are some basic templates provided, and accordingly the operational teams can add more data or customize the template provided.

Device Health Check

Devices	**CPU usage**	**Used memory**	**Memory usage**	**Overall status**
Storage A				
Storage B				
Firewall A				
Firewall B				
Firewall C				
DMZ A				
DMZ B				
DMZ C				
VPN				
DC core switch A				

DC core Switch B				
Router A				
Router B				
Database A				
Database B				

Application Health Check

Server name	Type of monitoring	Availability %	Performance %	Tools used	Owner	Actions taken	Status

Rack Health Check

Rack name	Total RU	Used RU	Max temperature	Current temperature	Weight	Status

Network Equipment Health Check

Network name	Asset name	IP address	Location	SPOC name	CPU Mhz	Memory	Brand name	Status

Bridge –ab-01	Bridge							
RAID-array-xs04-01	RAID array							
filestore-finance-02	Storage							

Data Center Energy End Use

Data Center End Use	**Electrical Demand (kW)**	**Estimated Annual Energy Use (kWh)**
IT Equipment		
Power Delivery Losses		
Supply Fans		
Cooling Compressors		
Pumps and drycooler, condenser, or tower fans		
Humidification		
Lighting		
Other		
Totals		

Energy Consumption Report

Month	**Demand (kW)**	**Power Factor (%)**	**Billed Demand (kW)**	**Electric Energy (kWh)**	**Electric Cost**	**Gas Use (Dth)**	**Gas Cost**
January							

February							
March							
April							
May							
June							
July							
August							
September							
October							
November							
December							
Total							

DC Asset Disposition

Asset disposition is an activity in ITAM (IT Asset Management) which focuses on disposing the obsolete and unwanted assets in a meticulous procedure such that no person can retrieve or access any kind of data or residual data from the assets and with the assets being destroyed in a safe and ecological manner.

DC asset disposition services generally involves

- On-site data erasure, hard drive crushing, and degaussing without any environmental hazards.
- Off-site destruction services which involves packaging, collecting, and transporting DC assets and decommissioning them without any data security risks and environmental hazards.
- Providing lockable storage bins to secure accumulated DC assets (like hard drives, networking devices, etc.) until destruction services are provided.

Disposal of a DC asset is not just a simple activity, it requires a secure, end to end process which involves degaussing and shredding.

Degaussing is the activity of totally erasing data by reducing or eliminating an unwanted magnetic field (information) stored on tape and disk media. A degausser is a machine that changes the magnetic domain (where the data is stored) of magnetic data storage devices.

When the degausser is applied to magnetic domains the information is scrambled into random patterns, making the data stored in the magnetic domain unreadable. The correct use of the appropriate degausser will ensure that information is no longer retrievable.

Shredding is activity which involves chopping, mincing, and shredding hard drives, cell phones, tapes, and other data storage devices into tiny, inaccessible fragments.

Key attributes to be maintained for asset disposal activities are:

- Owner/Name of the Requestor for disposal
- Equipment Type
- Department name
- Number of Items
- Manufacturer
- Model
- Serial Number
- Asset Number/Tag
- Reason for Disposal
- Disposal method (Donation/Reuse in other low priority business departments/Scrap)
- Date of Formatting
- Done By
- Validated By
- Authorized By
- Mode of Disposal

Template for Asset Disposal

Data Center Details

Data Center Name	Location	DC Manager	Telephone number

Asset Details

Asset no.	Serial no.	Asset description	Asset location	Owner name	Date of disposal	Hard disk formatted

Note: Key asset attributes (as mentioned in the previous page), can be populated in the above table as per the requirement

Disposal Approval

Name of Person Approving	Designation	Signature	Date

Questions to Be Answered Before Buying a Server

Investing and procuring on a server/data center equipment is a big investment which represents strategic decision and long-term commitment. Hence, here are some aspects that we have to focus on before purchasing a server.

- Understand what our budget is, accordingly you can consider classes of servers based on brand, size, and warranty details.
- Understand what is the purpose of the server?
- Understand what is the reliability of the server?
- Understand the space availability in your data center racks, and accordingly select the server with respect to the shape and size.
- Understand the storage and memory capacity needed for the server, also check on the scalability with respect to the future needs.
- What is the technology requirements like operating system, hardware configurations (RAM, HDD, Processor speed)
- Do you have a specific server type in mind? Blades, racks and tower servers, etc.?
- Are you replacing an existing server? If so, please have the make and model (or manufacturer part number) on hand when speaking to a member of the team.
- Do you have a rough estimate of how many users the server will need to support, as well as how many will need to use it at the same time?

- Do you know which applications will be running on the server? What hardware specs do the applications need to run?
- What is the redundancy available in the server (to avoid single point of failure) with respect to memory, hard disk, power supply, etc?
- Understand what are the hot swappable components available in the server?
- Check if the server supports fault tolerant technologies such as ECC (Error Correcting Code)?
- How much data will the server need to store, both now and in the future?
- What backup provision do you require for the data that will be stored on the server?
- Can you afford any downtime in the event of a disk or power supply failure?
- What level of support do you need?
- What kind of availability is needed?
- Do you need any special requirements with respect to compliance like PCI compliancy, HIPPAA, etc?
- Do you need a server to run applications, or just to store data?
- Does this server have ample and/or easily expandable storage capacity? Are easy upgrades available? is it scalable?
- Does this server make remote access to data simple or cumbersome?

Due Diligence Questions for Data Center Management

Due diligence in IT operations is an activity which conducts an investigation into operations; it ensures that you have access to all important information about the IT operations that you are going to manage. It enables you to assess the contracts, intellectual property, current operations, risks, issues, legal and regulatory compliance associated.

Due diligence questions for Data center management can be mentioned as:

- What are the customer points of concern with respect to Data center management?
- What is the scope of services for Data center management operations (like applications, infrastructure, etc.)?
- Are there any internal/external suppliers involved in supporting the Data center management operations?
- Is there a defined SLA for Data center management operations?
- What all regions do you support with respect to Data center management operations?
- Do you have defined Data center management process and policies?
- What are the KPIs and metrics defined for Data center management operations?

- Does your Data center management staff follow the process and policies? Or is there any resistance?
- Is there a defined organizational chart and roles and responsibilities defined?
- What are the different tools used for Data center management?
- Is the team well trained on tools?
- Is there any defined methodology for doing the risk and impact analysis?
- Is there a defined methodology for testing the changes?
- What are the different kinds of reports generated and shared with customer?
- What is the reason for outsourcing Data center operations?
- Have you ever outsourced the operations before? What were the issues you had with your previous supplier?
- How many users or host ports are currently required in the wiring closet, and how many will it require in the future? Should the switches be fixed or modular configuration?
- How many organizations or locations independently manage or operate your workload automation environment?

- How many hardware and people resources are required to operate your workload automation environment?
- How many different job scheduling solutions are currently in use in your various organizations or locations and in your overall environment?
- How many ports are available for end-user connectivity at the walls of the buildings?
- How many access switches are not located in wiring closets?
- What cabling is currently available in the wiring closet, and what cabling options exist for uplink connectivity?
- What data link layer performance does the node need?
- What level of redundancy is needed?
- What is the required link capacity to the Building Distribution layer switches?
- How will VLANs and STP be deployed? Will there be a single VLAN, or several VLANs per access switch? Will the VLANs on the switch be unique or spread across multiple switches?
- The latter design was common a few years ago, but today end-to-end VLANs (also called campus wide VLANs) are not desirable.
- Are additional features, such as port security, multicast traffic management, and QoS (such as traffic classification based on ports), required?

- How many devices will each Building Distribution switch handle?
- What type and level of redundancy are required?
- How many uplinks are needed?
- What speed do the uplinks need to be to the building core switches?
- What cabling is currently available in the wiring closet, and what cabling options exist for uplink connectivity?
- Are there any single points of failure?
- What level of availability does your business require?
- How much data is the business willing to lose in a catastrophic event?
- How much revenue will be lost or additional expenses incurred by an unscheduled system outage for different durations?
- What reputation risk or regulatory consequences are at stake for certain processing commitments?

Data Center Management for Consultants and Management

Improving Data Center Management

Define

The define phase should define the improvement opportunities, business case, goal statement, project plan, and roles and responsibilities.

Opportunity Statement – should define the issues in existing Data center management operations such as:

- Failures that are impacting the business
- Lowering cost of operations
- No process definition or inefficient process definition
- No defined procedures
- Growth is expected to exceed power, cooling, and space capacities

Business Case – should describe the benefits and opportunities of improving Data center management considering the following areas:

- What are the short term benefits and long term benefits of this project?
- What impacts will this new initiative have on other business units and employees (pros and cons)?
- Reason for initiating this improvement project should include:
  - Number of failures
  - SLA breaches
  - Penalizations involved
  - Non conformances observed in internal audits or external audits

Goal Statement – should be focusing on business goals and IT goals

- Should define what the improvement is, and what the team is aiming to accomplish? like cost reduction, SLA adherence, etc.
- What is the estimated time for delivery of results?
- What are the tangible result deliverables?

Project Plan – should show the milestones and time estimation for the activities involved in improving Data center management as a process and operations.

Roles and Responsibilities – should identify the required roles. Primary roles needed for change management can be defined as:

- Data center manager
- Facilities manager
- Operations lead

Define RACI model for clarity in roles and responsibilities.

Measure

The measure phase should capture as much as information about the current operations to understand how it is being operated and how well it is being operated. It also includes setting up the baselines as the basis for the improvement.

- Understand the IT services running in the organization.
- Understand the scope of services, and identify the assets. Create an inventory of hardware and software assets in the DC.

- Map applications to server/mainframes, database/platform, application dependencies, application security, etc.
- Identify the risks, cost assumptions, etc.
- Understand what are the teams, and which are facing maximum number of issues.
- Consolidate the list of all failures that have happened in the past by doing trend analysis.
- Identify number of changes that have happened and where there is maximum number of changes with respect to services.
- Understand the existing processes and procedures (for incident, problem, and change management).
- Identify who are the key contacts/SPOC's for different services.
- Understand the BIA.
- Understand what are the risk mitigation procedures and how risks are identified.
- Check with the staff on whether or not they have clarity and understanding on the processes and procedures.
- Interview the staff personnel to check what was good/bad/what was missing in the existing process and procedures.
- Perform baselines of the current change management operations.

Analyze

Analyze phase should identify the gaps and root causes for the discrepancies in DC operations. It also includes analyzing if the process collaterals processes, RACI matrix, SOPs, CSFs, KPIs, metrics, etc.

- Analyze what are the gaps in existing DC process and also procedures for operational teams, with respect to ISO/IEC20K standard.
- Analyze if the CSFs, KPIs, and metrics are aligned towards the IT goals.
- Analyze the existing metrics and SLAs, if they are feasible.
- Analyze if the BIA is performed correctly; identify the discrepancies if there are any.
- Analyze the risk mitigation methods and procedures and identify the gaps.
- Analyze if there any discrepancies in the criteria with respect to major and minor changes.
- Analyze what is the process and procedures for major incidents.
- Analyze how the root cause analysis is done on recurring incidents.
- Check if the staff is able to meet the defined metrics and SLAs.
- Identify the application to server/mainframe mapping.

- Identify the application to database/platform mapping.
- Identify the server to services and applications mapping.
- Identify the security requirements, applications usage, and its associated SLA's.
- Analyze if the DC staff is aware of the process and procedures?
- Identify and analyze the software assets (by business area, platform, SLA tier, etc.).
- Identify and analyze the number of servers.
- Identify and analyze the virtualization by types.
- Identify and analyze the server capacity and utilization by type (Ghz, Average %).
- Identify and analyze the storage capacity by type (Total/ used TB).
- Analyze the PUE, power density, rack density, rack space utilization, etc.
- Identify the flaws or misses in the existing process with respect to the business scenario by checking:
  - if the process is aligned to business objectives
  - if the process provides clarity in roles and responsibilities
  - If the process is clearly written without any ambiguity

- If the process is aligned to the best practices and standards
- If the process is repeatable

Note: Analysis on all the above activities can be done by brainstorming with the key stakeholders of DC management.

Improve

Improve phase aims at improving the current process and operations, with creative solutions to eliminate the root causes in order to fix and prevent process and operational discrepancies.

- Implement if any processes are missing (like implement software license management for managing the software licenses).
- Identify the root causes for all the failed changes (if they are less in number) or a sample of failed changes.
- Identify the root causes for all the changes that breached SLAs or a sample of breached SLAs.
- Involve all the key stakeholder from CM and other ITSM process and brainstorm how to improve the CM process and operations.
- List all the discussed ideas, and propose your ideas:
    - Redefining the CSFs, KPIs and metrics consulting SLM, as per the customer's requirements. Metrics should be defined in alignment to:
        - ✓ Customer's priorities, requirements, and acceptance criteria

    - ✓ Most important services

- Redefine the change management process ensuring:

    - ✓ There is a clear definition, demarcation and conditions on normal, standard and emergency change.
    - ✓ There is a clear definition on impact, urgency, and prioritization.
    - ✓ There is a clear definition of inputs, triggers, outputs and interfaces for the process.
    - ✓ There is a clear definition of change models for different services.
    - ✓ There is a change coordinators contacts list.
    - ✓ There are defined procedures for assessing the business risk and technical risks.
    - ✓ There is a defined test plan focusing on functionality, usability, performance, security, availability, etc. for every change request.
    - ✓ There is a defined procedure for the change manager with respect to daily job activities.
    - ✓ There are workflows/flowcharts that are easily understandable.

- Seek approvals from the Service Manager & Process owner on the redefined process collaterals.

- Ensure that all staff is aware on the usage of the change management tool with respect to change planning, collisions, impact analysis, etc.

- Define policies to conduct training sessions on process, knowledge, tool knowledge, etc.

- Define policies to assess the staff knowledge, conducting exams, quizzes, etc.
- Provide access to CMDB for all change management staff.

Control

Control phase defines the V&V (verification and validation) and continuous improvement activities that should be performed on change process and operations.

- Monitor the change management operations for the next two months, ensuring that all staff follows the defined processes and procedures.
- Check if the operational staff involved in change management is good to manage the number of changes requests occurring.
- Ensure that the change management staff is aware of the SLAs for all services.
- Check if the staff is able to meet the defined metrics and SLAs.
- Ensure all change requests are being tracked in the defined tool.
- Check the change requests which have breached the SLAs and perform the root cause analysis.
- Document the lessons learned.
- Update the process again if there are any changes in the business requirements.
- Prepare monthly reports (including accomplishments, targets met as per SLA, targets breached, total number of changes logged and implemented, successful changes, failed changes, failed changes that created incidents; they should include financial

information about the changes and incidents that are unresolved).

- Collect feedback from operational staff and customers.

Key Processes Needed for Data Center

Event Management

As per ITSM best practice frameworks, Event Management is a process which manages the lifecycle of events, which means it defines the sequence of activities for handling and managing events in IT infrastructure. Events here means any important occurrence like any failure, error message, etc.

Event management in Data center environment will play a very important role as it can:

- Provide real-time device failures within floor layout and rack front view for live status of data center operations
- Provide Power Usage Effectiveness (PUE) or Data Center infrastructure Efficiency (DCiE)
- Provide real time observation with respect to CRAC conditions
- Reduce costs by eliminating unnecessary servers which are taking the rack capacity.
- Identify underutilized servers to optimize IT capacity at the rack, row, and room level.

Incident Management

As per ITSM best practice frameworks, Incident management is a process which defines the process for managing the lifecycle of incidents, which means registering and categorizing, investigating and diagnosis, resolution and closure of incidents.

Incident management in Data center environment will play very important role as it will:

- Track and record every incident until the resolution and closure within defined SLA's.
- Identify major incidents and will follow special procedures to communicate, engage and resolve major incidents.
- Automate logging of incidents through the help of event management tools and DCIM tools.

Problem Management

As per ITSM best practice frameworks, Problem management is a process which defines the process for managing the lifecycle of problems which means registering and categorizing, conducting RCA, maintenance of known error database, identifying permanent fix and closure.

Problem management in Data center environment will play very important role as it will:

- Track and record every problem until resolution and closure
- Automation of RCA and identification of permanent fixes

Configuration Management

As per ITSM best practice frameworks, Configuration management is a process which defines the sequence of activities for managing the lifecycle of CI's, CI information, encompassing the information between CI's and its associated services, and auditing the CI's to ensure the information tracked and maintained is accurate.

Configuration management in Data center environment will play very important role as it can:

- Shows the consequences of device failure in data center infrastructure equipment for instant identification of critical business application impacts.
- Track CI addition, movement, and changes by date/time, owner and work orders for view, print and export.
- Increase accuracy and saves time by auto-associating discovered IT devices, based on properties such as make, model and IP address.
- Saves time and increases accuracy of the CMDB.

Change Management

As per ITSM best practice frameworks, Change management is a process which defines the sequence of activities for managing the lifecycle of a change request, which means registering and categorizing, assessing the impact and risk, change approval, change implementation etc.

Change management in Data center environment will play a very important role as it can:

- Schedule, assign and track the status of change requests to implement changes in the data center.
- Enables to know the past, future, and pending changes to assist with resource and workload balancing, avoiding scheduling conflicts.

- Facilitates scheduling of service and equipment maintenance to ensure visibility to all service plans and avoid scheduling conflicts.

Availability Management

As per ITSM best practice frameworks, Availability Management is a process which defines the sequence of activities for defining the availability configurations, monitoring the availability, analyzing availability issues, defining availability plan, assessing risks and their impact on availability, etc.

Availability management in Data center environment will play very important role as it can:

- Provide detailed view on availability for all the services in DC with respect to uptime, downtime, etc.
- Provides view on redundancy, fault tolerance, etc.
- Performs automatic service failure analysis, component failure impact analysis, fault tree analysis.

Capacity management

As per ITSM best practice frameworks, Capacity management is a process which defines the sequence of activities for defining the capacity and performance configurations, monitoring the performance, analyzing performance issues, defining capacity plan, assessing risks and their impact on performance, etc.

Capacity management in Data center environment will play very important role as it can:

- Provide an instant calculation of actual space, power and cooling consumption

- Provide an instant view on network port availability and floor and rack weight limits

IT Asset Disposal Management

IT Asset disposal management is a process which defines different procedures and sequence of activities for disposing the assets in a way that it generates ROI while disposing the assets, reduces IT management and burden, and generates no hazardous waste.

IT Asset disposal management in Data center environment will play very important role as it will:

- Cast off racks, rail kits, power cords, cabinets, etc.
- Remove obsolete equipment and supplies like transformers, circuit breakers, test equipment, reels of cable, spare parts, etc.
- Perform equipment recycling and disposition in an eco-friendly process that follows the strictest industry regulations and mandates.
- Ensures that all sensitive information has been completely destroyed prior to recycling.

IT Inventory management

IT Inventory management is a process which defines the process and procedures for tracking, storing and utilization of assets for day to day operations.

Inventory management in Data center environment will play a very important role as it will:

- Act as a repository that lists all hardware assets and shows the updated, detailed configuration and physical location of each workstation, server, accessories, etc.

- Track assets throughout their life cycle and will also enable you to view all hardware attributes, including CPU, BIOS, disks, devices, etc.

Project plan for DC setup

DC project plan is a formal document that should display the project activities along with timelines. It should depict what, when, by whom, etc.

Here is one such template which gives you a comprehensive description of a DC project, which can be modified as per the requirements:

Activities	**Start date**	**End date**	**Responsibility**
Nominate DC Head			
Send information to organizational departments like Networks & Services & HR about DC Setup			
Prepare Draft Services Catalog			
Prepare Draft Infrastructure Plan			
Review Draft Service Agreement with the sales/pre-sales team			
Facilitate legal and managerial review of Draft Service Agreement			

Signing of Master Service Agreement			
Request for meeting with the customer			
Review / Refine Services Catalog			
Review / Update DC Setting up Plan			
Gather Applications Inventory			
Analyze Applications portfolio and prioritize the services			
Analyze Physical Infrastructure requirements			
Analyze Office Space requirements			
Analyze Telecom Links requirements			
Analyze Physical & Data security requirements			
Analyze Infrastructure Maintenance requirements			

Analyze H/W & S/W requirements			
Prepare DC Operations Manual			
Identify core DC Teams and define preliminary Roles & Responsibilities			
Develop Reporting Templates			
Prepare DC Performance Metrics and Measures			
Define ITSM processes for DC (Incident, Problem, Change management, etc.)			
Define Business continuity & Disaster recovery			
Test & certify infrastructure			
Prepare Knowledge Transition Plan			

Knowledge transfer to core team			
Finalize Performance & Metrics requirement			
Finalize Physical Security Management			
Finalize NDA format			
Knowledge Transfer acceptance/ Signoff			
Evaluate DC Metrics			
Conduct audit for security			
Conduct metrics audit			

High Level View of DC Setup

DC setup is a project which involves a number of activities and any discrepancies, delays, or any issues in the project which might lead to huge financial losses, reduced quality in delivery of services, higher costs in setting up the DC, project schedule overrun, etc. leading to customer dissatisfaction.

Hence, DC projects should follow a process which involves: Planning → Designing → Acquiring → Implementing → Maintaining and operating

Planning:

Planning lays the foundation for any project, and also the same case with the DC projects; this is the phase where the management has to play a vital role in spending time with requirements to collect all the requirements, assigning the right SME's to analyze all the issues and the feasibility of the project. Here are some important activities that have to be performed in planning phase:

- Understanding the requirements and business goals of the customer
- Understanding the preferences and constraints from the customer (e.g. overhead power cabling, security cameras to see every inch of the data center space, wide IT racks to provide more cable space, etc.)
- Documenting the understood requirements and goals
- Signing off by the customer on the agreed requirements
- Defining the money planned for the capital costs of the project
- Defining the Data center site

Based on the gathered requirements, a specific country and city is shortlisted (considering the risk of natural disasters, power supply, weather, transportation, availability of skilled human resources, etc.)

- Defining the roles needed and appointing the people necessary

Designing:

Design is another crucial stage where all the technical work is prototyped with respect to integrated infrastructure systems, ups & power distribution, generator & ats, battery systems, computer room air conditioning, cooling distribution, heat rejection equipment, raised flooring, fire detection systems, pre-action sprinkler systems, environmental monitoring systems, leak detection equipment, cctv & ip-camera systems, access control systems, racks, cabinets & accessories, structured cabling, cable management, noc, console & technical furniture, network & security infrastructure, server & storage systems, etc.

Here are some important activities that have to be performed in design phase:

- Description of electrical layout, ramping up to the maximum power requirement, incorporating uncertainty.
- Description of the resources like servers, networking devices, storage, cables, etc.
- Model of the floor space (Total Space Requirements and Power Density for Data Centers), network layout.
- Definition of staging area, storage area, cooling systems.

- Understanding and defining the level of system availability to be achieved.
- Understanding the capacity and performance (Maximum IT load) that the data center physical infrastructure can support.
- Defining the server infrastructure, virtual infrastructure, storage, application communication, etc.

Acquiring:

This is the stage which involves selection of the suppliers, investigating the products/services, purchasing the equipment, stocking the equipment, site coordination, etc.

Implementing:

This is another important stage which involves installation of racks, physical mounting of hardware, Network switching and routing, SAN/NAS/Backup, Firewalls, security zones, enterprise shared services (authentication and authorization, antivirus and anti-spyware, security services), core infrastructure and shared services (Intranet, extranet, internet & VPN connectivity testing) and testing of high availability/failover testing, performance/stress testing, ethical hacking/vulnerability assessment/penetration testing.

Maintaining and operating:

This is the final stage where the customer gets to see the value of the Data center operations. Maintaining and operating is to coordinate and carry out the activities required for ongoing management of the technology to deliver and manage services at agreed levels to business users and customers. Data center maintenance and operations activities can be managed by the ITIL Service operations process area, through the help of:

- Event Management Process: A process to support Data Center Events, Event logs, Changes including Policies, Principles, Process Activities, Methods, Triggers, Inputs, and Outputs.
- Incident Management Process: Process to Support Data Center Incidents such as outages, equipment loss including Policies, Principles, Triggers, and Techniques for recovery.
- Problem Management Process: Handling of Identified Problems or Issues within the data center including root cause analysis, comprehensive fixes, improvements, and knowledge library inputs for future problem resolution.
- Request Fulfillment Process: Approach to handle end user, organization requests for moves, adds, or changes to infrastructure within the data center.
- Access Management Process: Process for server, application, database, and physical access to the data center and its assets for end users.
- Service Desk Function: Provides a support organization with rules and responsibilities for service support to end users.

Maintenance and operational activities primarily include:

- Develop an annual preventive maintenance schedule
- Coordination and tracking of all service visits
- Physical site inspection
- Resolution of service issues
- Data center cleaning
- Maintenance of air conditioning equipment, Racks, HVAC, UPS & Batteries
- Power & data cable physical inspection
- Maintenance of generator and automatic transfer switch equipment, fire suppression & detection equipment, sprinkler systems

Aerial Overview on the DC Setup Activities

- Definition of the roles and responsibilities for the project.
- Nomination of the DC head.
- Gather requirements from the customer about the DC.
- Get the approval from the management/customer to setup the DC.
- Draft the DC setup plan with respect to infrastructure systems, power distribution, computer room air conditioning, fire systems, CCTV & IP-camera systems, access control systems, cable management, network & security infrastructure, server & storage systems.

- Draft the infrastructure, service offerings, and transition plan.
- Send the information to the respective stakeholders like network, HR, finance, legal, quality, etc.
- Get the MSA signed by the customer/ management.
- Draft the list of services that will be offered.
- Draft the infrastructure resources and human resources needed for running this project.
- Draft the DC operations manual & processes needed.
- Get the drafted list of services, documentation, infrastructure resources and human resources - approved from the customer.
- Initiate the project kick off meeting.
- DC head will revise the plans and documentation collaterals as per the demands of the management/ customer.
- DC head will revise the infrastructure resources and human resources with respect to the plans defined.
- Once the infrastructure resources are procured they are updated in the inventory repositories.
- Test the infrastructure resources and its associated documentation.
- Execute the operations.
- Define a knowledge management system which should accommodate the knowledge articles with respect to day to day operations.
- Decide and finalize the DC metrics, KPI's and auditing procedures with quality team.
- Conduct regular security and metrics audit.

- Identify improvement initiatives and consolidate them in an improvement plan.

High Level View of ODC Setup

The ODC setup process can be subdivided into five phases as planning, requirements gathering, implementation, transition, and audit and improvement. The below mentioned activities can be split into these five areas.

Aerial Overview on the ODC Setup Activities

- Definition of the roles and responsibilities for the project.
- Nomination of the ODC head.
- Gather requirements from the customer about the ODC.
- Get the approval from the customer to setup the ODC.
- Draft the ODC setup plan.
- Draft the infrastructure, service offerings, and transition plan.
- Send the information to the respective stakeholders like network, HR, finance, legal, quality, etc.
- Get the MSA signed by the customer.
- Draft the list of services that will be offered.
- Draft the infrastructure resources and human resources needed for running this project.
- Draft the ODC operations manual.
- Get the drafted list of services, documentation, infrastructure resources, and human resources - approved from the customer.

- Initiate the project kick off meeting.
- Service provider will revise the plan and documentation collaterals as per the demands of the customer.
- Service provider will create the infrastructure resources and human resources with respect to the plans defined.
- Once the infrastructure resources are procured they are updated in the inventory repositories.
- Test the infrastructure resources and its associated documentation.
- Get the approval from customer on knowledge transition. If the knowledge transfer is performed by customer, then the service provider gets the sign off from the client.
- Execute the knowledge transition.
- Execute the operations.
- Define a knowledge management system which should accommodate the knowledge articles with respect to day to day operations.
- Decide and finalize the ODC metrics, KPI's and auditing procedures with quality team.
- Conduct regular security and metrics audit.
- Identify improvement initiatives and consolidate them in an improvement plan.

Risks in Data Center Management

Risks assessment and management is very crucial activity and process that should be must in data center management. Risk assessment and management helps in:

- Determining which systems are not matching with business requirements
- Identifying vulnerabilities and threats
- Identifying single points of failures
- Identifying and ranking the IT assets based on criticality to the business
- Ranking the vulnerabilities, threats, and risks

Some of the common risks that are found in DC management are:

- End-user access and security
- Server failures Expected or unexpected downtime while migration
- Schedule and budget overrun while migration
- Inadequate baseline discovery while migration
- Software licensing and compliance
- IP addressing and hardcoded IP addresses
- Undetected smoke that can lead to fire accidents
- Lack of communication between the staff
- Discrepancies in inventory management
- Delays in procuring assets

- Network connection failures
- External hackers
- Inefficient management of Data center keys
- Outsourcing Data centers - because the company's critical business information is run and stored externally

Automation in Data Center Management

Today, most of the organizations are having challenges in performing simple routine tasks in DC, which takes time and unnecessary human efforts; hence, Data center automation is a vital step for achieving the business results in an effective way. DC automation will enable to automate IT processes, procedures across computing, network, and storage layers in all environments. Also automation of the data center will reap benefits of increased efficiency in operations, decreased cost of management and power, etc.

Data center automation can help the team to:

- Simplify management of ever-growing routine IT tasks.
- Distribute applications intelligently across multiple resources so that the capacity and performance needs of the application are met by the appropriate hardware.
- Identify the possible locations where the new servers can fit in.
- Filter the racks by space, power, and network capacity, displaying a colored-coded map of suggested destinations giving information on availability space, available kW, connections to redundant PDUs, and access to secondary power sources.
- Keep the environment up to date.
- Automate deployments, and update patches on servers according to policy.
- Audit live configurations.

- Compare configurations to each other, whether live or historical snapshots.
- Verify pre-requisites before deployment.
- Estimate the impact of a change.
- Evaluate the risk of a change.
- Send automatic notifications, reminders, and emails to the stakeholders.
- Identify the CI dependencies.
- Base-line a CI or an IT service.

Cables Management in Data Centers

Cable management in Data center

Cable management is one of the most important aspects of data center design. Cables are the lifelines for the data center operations, as they carry the data and electricity to data center infrastructure.

Cable management in data center can be categorized into three sub processes: planning the cable infrastructure, selecting cabling components, and implementing cabling and testing cables.

Planning the Cabling Infrastructure:

Planning the cabling infrastructure involves the strategic activities providing direction for cable management in data centers. It involves activities like:

- Understanding the cable types (copper, fiber, etc.), counts, approximate routed distances, equipment ports needed.
- Defining a color convention to identify cables.
- Establishing a naming scheme considering building number/name, room number, rack number, patch panel, workstation outlet, port.

Selecting Cabling Components

Selection of the cables and cable components should be in adherence with compliance to EIA/TIA/IEEE standards. Other factors to be remembered while selecting cables are:

- Avoiding cheap cables
- Buying thicker cables based on the requirement
- Color selection based on the policies defined
- Compatibility with the racks
- Space for labeling on the front of the panel

Implementing the Cabling and Testing Cables

Implementing and testing cables is another important activity that should be done through the help of an expert. Implementing cables will happen under a raised floor, overhead, or both. Here, we need to ensure that the cables are individually separated, remembering the weight of one cable is not affecting the other light weight (fiber) cables.

Testing cables activity should be in place right from installation, even in the case of relocation or termination of cables accompanied by preparing test reports.

Implementing the cables should be done ensuring:

- There is a scope for future growth
- Cables are not affected due to bend stress
- Plenum rated cables are used wherever needed

Some tips for cable management

1. Avoid over-bundling cables.
2. Keep fiber and copper cables separate.
3. Don't stress the cables by applying additional twists, pulling, and bending.
4. Avoid cables near equipment that generates EMI.
5. Avoid exposing cables to direct sunlight.
6. Label cables with their destination at every termination point.
7. Maintain the cabling documentation, labeling, and logical/physical cabling diagrams.

Data Center Assets Lifecycle for Management

Data center assets lifecycle would also be the same like any other IT asset in an organization.

Plan: This step involves understanding the requirements from the internal/external customer with respect to server configurations, power usage requirements, air conditioning requirements, ports, cables, etc. After consolidating the requirements, it has to be brainstormed with SME's. Then after approval, it moves to the next step requisition.

Requisition

This step is triggered by the IT team sending a request for new hardware/server. Then the inventory team checks if there are any assets available in the inventory with the specified configuration and will decide to purchase a new asset or use any purchased asset in the inventory.

Procurement

This step involves procurement team's work in obtaining the quotation from different suppliers and choosing the best one ensuring value for money.

Acquisition and Stock

This step involves acquiring the assets from the suppliers, verifying whether if we have received the right assets as per the requirements with respect to CPU, memory, power supply, fans, storage, etc. and stocking it in the inventory.

Discovery

This step involves discovering the HW configuration of each device, and the server as a whole so that it can be deployed correctly.

Qualification

A new server should be initially quarantined by firewalling and also assurance checks are to be made at three aspects: authentication, scanning to detect malware, and compliance checks to ensure that it suffices the IT policies.

Bare Metal Provisioning

This step prepares the server for installation and involves a variety of tasks such as HW partitioning, configuration, tuning, and loading basic system software.

Service Provisioning

This step would assign the server to different project or services with necessary application software, network/storage access, etc. so that they can provide intended services.

Monitoring

This refers to constant monitoring of vital parameters of the server and taking suitable actions. Monitoring data from various HW and SW elements would typically involve filtering, storage and fusion in order to detect and resolve performance problems, minimize power consumption, determine security attacks, etc.

Remediation

This refers to activities related to fault detection/diagnosis, security related quarantine, repair, upgrade and replacement. Remediation may be required while the infrastructure equipment is in use and thus may interfere with the service.

Retirement and Disposal

This refers to activities related to the disposal of retired and obsolete assets as per the organizational, legal, and environmental requirements.

Data center equipment will be disposed when the equipment meets the below mentioned criteria:

a) The equipment is not required for the organization due to changed business goals and end products.
b) The equipment has become inefficient to operations.
c) The equipment has negative impact on service delivery.
d) The equipment become hazardous to use.
e) The equipment has met or crossed the expiration date.

Data center equipment disposal can be done based on the asset's functionality, associated financial value, asset utilization, asset average lifetime, and other factors. Some of the well-known methods are:

a) Reassignment to less business critical operations like using in libraries, reception, lobbies, etc.

b) Selling in a public forum.
c) Donations to schools or orphanages.
d) Discarding as scrap after removing the toxic materials considering SHER (Safety, Health, Environment, and Risk) issues.

Data Center Migration

A Data center migration is the activity which involves transferring and deploying an existing data center environment to another data center operating environment. Data center migrations require a well-designed strategy and comprehensive planning throughout all phases of the project which must be executed without impacting business operations, SLA's, availability, performance, and security requirements.

Data center migration is normally triggered for the following reasons like IT cost cuttings, regulatory requirements, technology complexity, etc. Below is a more detailed explanation of the reasons:

- Acquisitions, mergers, demergers of organizations leading to a new Data center unit.
- When the Data center is fully utilized and is out of capacity.
- When the Data center's assets have reached the EOL support or when the assets are increasing more costs like high power consumption, etc.
- When the Data centers are in a risky zone where there is more number of chances for natural disasters like earthquakes, floods, etc.

Approach for Successful Data Center Migration

Successful Data center migration should definitely follow these below mentioned steps:

Analyze and Assess → Plan → Design → Test → Implement & Operate

Analyze and Assess

This stage involves gathering, analyzing and assessing the inventory information driving the overall migration strategy. It is an activity which checks the organization readiness and makes a baseline of the assets. It also includes performing a check on inventory of all virtual assets that will be handled. The objective of this phase is to evaluate the accuracy and completeness of processes, procedures and documentation; the focus areas include:

- Support structure and architecture
- SLA's
- Capacity planning and performance management
- Monitoring and management
- Resource availability
- Industry regulations
- Logistics
- Stakeholders
- IT Service continuity plans

And this information is gathered based on source Data center environment and target Data center environment.

Plan

Based on the as-is analysis and assessments made in the previous stage; here in this stage, a strategy and plans are made with respect to:

- Scope and size of the project.
- Plan and allocate resources and budget.

- Business requirements.
- Governance structure.
- Identify the project plan identifying the critical path and key resources, milestones, etc.
- Goals and objectives.
- Roles and responsibilities.
- Best practices for migration.

Design

Designing the Data center infrastructure involves activities like:

- Data center site selection – It involves considering environmental conditions, wide area network, power, construction environment, fiber network connectivity, transportation modes, land acquisition issues, construction costs, government laws, taxation, etc.
- Data center power management
- Data center cooling
- Data center migration facility
- Data center server infrastructure
- Data center shared services

Test

Testing the Data center involves activities like:

- Pre-migration testing – Verification of scope with respect to source systems and testing source and target system connections from the migration platform.

- Post-migration testing – After migration, another end to end testing is performed.
- User acceptance testing – User acceptance testing provides an opportunity for the user community to interact with legacy data in the destination system prior to production release, and most often, this is the first such opportunity for the users.

Implement and Operate

The Implementation phase comprises of the key activities like:

- Knowledge transfer from customer/vendor to gain complete knowledge of the applications, tools, environment, and processes.
- Executing the day to day IT operations.
- Reviewing the project status and communicating it to the management and customers.

Data Center Site Selection

Data center site selection is one key success for stable and continuous business operations. There are some important and critical factors to consider before choosing a site for a new data center facility as mentioned below:

- Electric Power: Power is typically one of the highest on-going costs of operating a facility; understanding the vulnerabilities and reliability of the power infrastructure components is an essential part of site selection. It requires good availability, economic cost, and redundancy for Data centers to run equipment. Choosing a location with desirable electrical rates is an important factor. Clean power is also a key criterion in the site selection process. Data center owners/operators seek reliable, efficient renewable energy options such as hydro, air, wind and, to a lesser degree, solar. Fuel cells are also emerging as a viable power option.

- Risk of Natural Disasters or Weather Events: An area that is prone to natural hazards may not be a suitable building location. Seismic events, floods, tornadoes, hurricanes, and volcanoes can all place the data center as well as suppliers of power and other services at risk. Weather events and natural disasters have the potential to interrupt operations and are of great concern when locating data centers.

- Weather: Climate can significantly affect the efficiency of data center cooling, which in turn affects sustaining costs. Temperate and climates are a growing subset of the environmental factor thus driving down PUE (power usage efficiency) - a major consideration in data centers today.

- Land & Construction Costs and Quality - The cost of the land is a major criterion. It is better to choose sites with an appropriate cost per square meter (acre) and prefer sites with single owners because multiple owners can slow negotiations and purchasing activities. Also it needs availability of qualified construction labor, such as framers, concrete workers, electricians, and plumbers, contribute to a successful construction project.

- Quality of Life - Data center employees are highly educated and enjoy higher than average annual salaries, and quality-of-life factors are an important consideration in attracting and retaining these employees.

- Labor Costs and Availability: Availability of technical labor, though data centers typically employ fewer than 100 employees it is very much essential that the labor is skilled and within reasonable commuting distance of the Data center. IT personnel, such as network staff, system administrators, data center architects, security staff, facility technicians and other maintenance staff (maintaining chillers, UPS, power distribution equipment) help keep the data center running after it is built.

- Availability of Telecommunications Infrastructure: Various types of media, such as fiber, copper, satellite, and microwave, are used in WAN connectivity. Understanding latency limits and the location of preferred fiber providers and telecommunication points-of-presence in advance of identifying sites increases the overall success of the site selection process.

- Tax and Regulations: The local, regional, and national taxes, regulations, and incentives can affect virtually every consideration for choosing a Data center location. Taxes may include sales, personal property and real estate. Regulations can be enforced at the local, regional, and national levels. Regulations that should be considered when choosing a data center site include land zoning, fuel storage and emissions, etc.

- Transportation: Proximity to prime corridors can be viewed as favorable condition as access to major transportation modes, such as airports, rail lines and interstate roadways, creates the potential for easy transportation link.

- Proximity to Water Resources: Water is a key resource for data centers because evaporation is often the most cost efficient cooling method.

Parameters for Data Center Site selection are:

Parameter	Availability condition	Remarks
Risk of natural disasters		
Weather		
Land & construction costs		
Labor costs & availability		
Transportation feasibility		
Power Reliability		
Power Quality		
Alternative Energy Options		
Electricity Rate Price		
Power Infrastructure Cost		
Power Infrastructure Timing		
Power Available Capacity		
Percentage of Utility Load		
Utility Financial Stability		
Privatized/Deregulated/Monopoly		
Power Customer Service		
Power State-of-the-Art Equipment		
Future Stability of Power Resource		
Power Line Specifics		
Water Reliability		
Water Infrastructure Cost		
Water Infrastructure Timing		
Sewer Reliability		
Sewer Infrastructure Cost		
Sewer Infrastructure Timing		
Natural Gas Reliability and Capacity		
Natural Gas Rate Price		

Auditing Data Center

The objective of audit is to ascertain the effectiveness of existing policies and procedures related to the administration and control of data center operations. Also, it is to determine the adequacy of controls over the related processes, and ensure compliance with relevant governmental regulations.

Key focus areas of the audit are to ensure:

- Data center policies and procedures are defined, documented, and communicated for all key functions.
- Council systems are secured to prevent unauthorized access (including 3rd party access).
- Access to the Data center is monitored and reviewed, and access rights are periodically reviewed.
- Data is backed up from servers held at the civic Data center.
- Data transferred off site is secured at all times and appropriate controls are in place to monitor the location of the data.
- Environmental controls are present to protect the servers from fire, electrical and water damage.
- Capacity for the Data center is adequate for the server rooms equipment and storage needs.
- Environmental equipment is routinely maintained in line with manufacturer recommended schedules.
- Backup electricity supplies are in place to ensure systems and services are not affected in the event of a power outage.
- Background checks are performed on all employees who have physical access to the data center.

- Data center room should have a security alarm system.
- There are procedures to monitor and review electronic key card access system reports that record the time, dates, and names of employees entering and exiting the data center room.
- Data center operations staff should be trained on emergency evacuation procedures, fire safety, and the use of fire extinguishers.
- Service Continuity Plans should be updated, tested & documented.
- Location of the primary Data center is away from risks like fire and explosion hazards, water related accidents, etc.
- There is an updated inventory of equipment, systems, and data residing in the data center.
- There is coordination with all agencies that have hosted systems in the data center to rank the systems' criticality and establishing a priority for the order in which systems will be brought back up.
- There is evaluation of existing threats to the data center including the potential impact or harm.
- There is a cost analysis conducted with implementing or improving controls.
- Is checking on electrical equipment happening on a regular basis for capacity planning?
- Energy savings plan is documented and rewarded.
- Energy savings plan is reviewed regularly and corrective action is taken to address failures.
- Energy usage is continuously monitored to determine peak and low energy demands.

Sample Audit Report Template

S no	Priority	Findings	Risk arising consequence	Recommendation	Responsibility and Timescale
Policies and Procedures					
1	L	Lack of archival policies & procedures	Increased risk that archival is not done appropriately	We recommend that the operations manual is including with archival procedures	Operations lead -End August 2016
Fire suppression system					
2	H	There is no fire suppression system	Increased risk that data center operations might stop and huge financial loss	We recommend to have fire suppression system asap	Operations lead -First week February 2016
Access logging					

3	M	As per the policy access to the Data centers should be logged	Unauthorized access to the Data centers room might lead into accidents like loss of data or interruption of services	When non-authorized staff is entering the room they should be accompanied by the project staff member	Operations lead -End March 2016

L – Low, H – High, M - Medium

Choosing a Data Center Partner

Choosing a Data Center Partner

Data center partner is a service provider who owns and takes the accountability of managing the customer's servers or services.

To choose a Data center partner, an organization must first understand:

- What are the organization's business and IT requirements?
- What is the organization's growth plan?
- What is the Must, Should, Could, and Would requirements?

Checklist for Evaluating a Data Center Partner

- Enquire on the company with respect to references and make sure that you are getting a good match on human resources experience and qualifications?
- Does vendor have financial stability? What is his core business (Is it only Data centers or anything else)?
- Are there any ownership changes in the last few years?
- What is the seismic zone of the Data center partner's location?
- Were there any floods in the last few years and what is the flood zone rating?
- Does the vendor have the infrastructure capacity to support your applications?
- What is its current capacity for bandwidth, computing, and storage resources?

- What is the vendor's plan for expansion of resources, and what reserve threshold triggers an expansion?
- How many sites does the vendor operate and what capacity resources are available at these sites?
- Does the vendor operate its own facilities or is its infrastructure collocated in another vendor's facility?
- Does the vendor provide the engineering support that is needed to design and effectively operate your solution within your performance goals?
- What proactive monitoring is in place for performance issues and who is notified?
- What types of resources are available on-demand? DBA, Network, Security Engineers?
- Will the vendor provide unfettered access to quality system documentation?
- Does the vendor make training and maintenance documents available?
- Will the vendor support audits by your clients?
- Will the vendor share disaster recovery plans?
- Does the vendor have enough power density to support both current & future technologies?
- Does the vendor support flexible master service agreement (MSA) and service level agreements (SLAs)?
- Do they have air conditioning and UPS & generator redundancy configuration?
- Do they have support from multiple network providers?
- Do they have zoned fire protection system and fire suppression systems?
- Does the vendor conduct scheduled proactive battery testing?
- Does the vendor conduct scheduled proactive generator testing?

- Does the vendor conduct scheduled proactive rack power testing?
- Does the vendor have locked cabinets and cages?
- Does the vendor have biometric hand scanners?
- Does the vendor have escorted access to Data center floor?
- Does the vendor currently support clients that are within a FDA or HIPAA regulated vertical?
- Does the vendor maintain an active quality system that can flow through to the client?
- Are clients notified of infrastructure and operational changes at the Data center infrastructure level?
- Does the vendor maintain change management and quality management duties at the client solution level?
- What is the audit history of the vendor? Have they been audited by a third party for a relevant regulatory structure and will they provide audit support for you?
- What is the SLA achievement of the vendor in the last three years?

Checklist for Effective Server Operations and Maintenance

Simple maintenance tasks can prevent disasters in a Data center and server environment. By keeping track of certain information IT team can identify issues before they become failures.

- Do you track and manage the server information, and do you have a defined standard server names?
- Is this a Physical or Virtual server?
- Is this a physical to virtual conversion of an existing server?
- What applications will run on the server, and what operating system, processors (GHz), memory (RAM), hard disk space (gigabytes) does the application require?
- Is the server placed in a secure office or preferably a dedicated room, where only authorized people can access the server?
- Is the booting from diskette, optical drives, & USB devices disabled?
- Are the unused hardware device ports disabled?
- Have all the services been patched to current levels?
- Has a viable plan been designed to maintain the server properly (including consistent regular patching of the OS and all applications)?
- Is the server adhering to organization's password policy?
- Has the server been correctly labeled?
- Does the server require SAN connectivity [Y/N]?
- If yes, please state if a LUN has been allocated [Y/N].

- If yes, please state the disk space required (FC or SATA).
- If yes, please state whether the server is to be mirrored.
- What VLAN is to be assigned?
- Define the inter-VLAN access required.
- Does the server need to be backed up?
- If this is a rebuild of an existing server, has the application data been backed up to a location off the server?
- Is the server plugged in to an Uninterruptible Power Supply (UPS)?
- Is the server adhering to the 'Installation and Patching' guidelines?
- Is the server having an appropriate name and a fixed IP address?
- Is the server having an appropriate VLAN?
- Do all the servers have access logging enabled?
- Are the logs checked regularly (at least weekly) for unusual access attempts?
- Do you make backups of important data and store the tapes in a fireproof safe and preferably not in the same location as the server?
- Do you have redundancy for all critical servers?
- Do you check for latest service packs and hotfixes?
- Are the admin passwords having a minimum of 15 alphanumeric and special characters? Has the OS been properly installed and configured and all relevant security patches and anti-virus software applied?
- Have all "network application" services not essential to the prime function of the server have been disabled (HTTP, telnet, FTP, SMTP, DNS, etc.)?
- Is enough disk space available?

- Are backups restorable and how much time does it take to restore?
- Are all hard disks in the RAID working? Check RAID alarms.
- Is the server load on acceptable levels with respect to (CPU/ RAM/ IO)? Check for hardware errors.
- Are there any no old accounts on the server which should have been removed but weren't?

Initial Assessment for Data Center Migration

During the initial assessment phase all services that have to be supported and are in scope will be analyzed and categorized according to criticality, complexity, and time to move. Here are some of the factors which have to be identified and analyzed:

- Determine Owned/Hosted Servers – check and understand which systems are owned and which are hosted.
- Determine Owned/Hosted Network Equipment - check and understand which systems are owned and which are hosted.
- Determine Owned Software Licensing – check and understand which licenses are owned, leased and etc.
- Determine Amount of Hosted Data – check and understand the amount of user data, facility shared data, globally shared data, and application data.
- Determine Transfer of Licensing – check and understand the account numbers and licensing transfer methods.
- Determine Server Migration – check and understand the schedule for migration of servers.
- Determine Network Equipment Migration – check and understand the schedule for migrating routers.
- Determine Data Transfer Method – check and understand the method of data transfer.
- Determine Number of Expected Users – check and understand the active user count for current number users to be used for licensing requirements and load balancing.

- Determine Number of Expected Services – check and understand the number of services to be provided.
- Determine Application Dependencies – check and understand the software requirements of each application listing dependencies, ports and configurations.
- Determine Number of Expected Servers – check and understand the number of servers at hosting provider, discuss load limits and consult application list for consolidation possibilities.
- Determine Network Requirements.
- Determine Rack/Space Requirements – check and understand the server requirements and number of VM hosts required.
- Determine Power Requirements – check and understand the number of host servers, storage arrays, and switches.
- Determine Cooling Requirements – check and understand the racked servers and space.

Business Continuity in Data Center

Business continuity is a process that defines the procedures for continuation of critical business processes when a disaster occurs. This plan should define the procedures for preparation, testing, and maintenance to recover normal processing.

Why BC in Data Centers?

BC in Data centers is essential as its going to bring a state of readiness for responding to a prolonged disruption in DC services caused by power failures, system failures, natural disasters, or any other unforeseen circumstances.

Disasters can be classified into three types

Disaster type	**Examples**
Natural	Earthquake, Tsunami, Tornadoes, etc.
Accidents	Fire, power outage, telecommunications outage, etc.
Intentional acts	Hacking, theft, terrorism, etc.

Fire Accident

Fire accidents can be considered as one of the critical disaster which can disrupt the DC services. Fire accidents can be caused due to various reasons like:

- short circuits
- smoking

- defective appliances
- inflammables

Controlling fire accidents can be done by setting up corrective and preventive measures like:

- Defining special procedures to address evacuation, medical help, and firefighting.
- Fire drills to be organized at regular intervals.
- Smoke detection systems and fire alarms should be available.

Power Outage

Power outages can also be considered as one of the critical disasters and one of the frequently occurring issues which we normally hear in DC operations.

Power outages can be caused due to reasons like:

- Short circuits
- Cable faults
- Grid fault
- UPS short

Controlling the power outages can be done by setting up corrective and preventive measures like:

- Performing routine maintenance on the electrical wiring
- Monitoring diesel consumption
- Automatic shutdown of lights and air conditioners when the employees are not present in office for long time

Telecommunications Outage

Telecommunications outages can also be considered as one of the critical disaster and one of the frequently occurring issue which we normally hear in DC operations.

Telecommunications outages can be caused due to reasons like:

- Physical damage to network components
- Disruption in supporting network infrastructure
- Network congestion

Controlling the telecommunications outages can be done by setting up corrective and preventive measures like:

- Conducting audits to determine if facilities can meet power demands and ward off problems.
- Paying particular attention to hardware that is approaching the end of life (EoL).
- Monitoring systems in real time 24/7.
- Managing proactive upgrades of equipment approaching EoL.
- Providing redundancy.
- Conducting a simple audit of your organization's underlying network to identify where such conditions exist.

Earthquakes/Tsunamis/Tornadoes

Earthquakes/Tsunamis/Tornadoes can be considered as one of the critical disaster which can disrupt the DC services.

Though they are heard of very infrequently, they cause maximum damage.

Controlling the natural disasters outages can be done by setting up corrective and preventive measures like:

- Equipping data centers with seismic isolation technology using shock absorbers at floor level and rack level.
- Starting uninterruptible power supplies and diesel-powered generators when there is a power outage.
- Selection of location which is free from natural disasters by looking at the historical events.
- By running dual centers, even if one data center suffers any natural disaster damage, the other one can continue operations.
- Building Data center based on industry best practices like ANSI/BICSI 002-2011.

Hackers

Hacking can be considered as one of the critical disaster and one of the frequently occurring issues which we normally hear in DC operations.

Hacking DC services can be caused due to reasons like:

Physical attacks like crawling through void spaces in the data center walls, lock-picking the door, "tailgating" into the building, etc.

Checklist for Data Center Staff

- Are we effectively measuring power utilization in our data center(s) today?
- What's our overall efficiency rating (PUE, DCIE, and CADE), and how does that compare to the industry norm?
- Are we able to accurately determine which equipment is being utilized the least and/or is the least efficient?
- Can we determine energy consumption by application, transaction, or department, and does it make sense to allocate those costs accordingly?
- How much is this year's technology refresh cycle expected to improve overall data center efficiency and capacity?
- What gains did we realize (or do we expect to realize) from consolidation and virtualization of server and storage resources?
- Can we optimize those gains with "what-if" analyses of all available options or through other measures?
- Where are we most likely to experience a serious limitation, and are we at risk of having a data center exceed its available capacity?
- How can we minimize energy consumption (and costs) in our data center(s)?
- What will be the return on investment for improving overall efficiency in our data center(s)?

Checklist for DC Migration

Checklists for DC migration will be helpful to consolidate the complete details of a source data center and to plan the requirements for target Data center. Here are some questions that can be mentioned as below:

- What is the approximate cost of new servers, software and other hardware?
- What networking space is occupied in source data center?
- Who are the key contacts for an application (Application manager, Delivery Manager, System Administrator, Architects, etc.)?
- What is the number and type of hardware devices?
- What is the number and type of operating systems, business and enterprise software services deployed?
- What is the existing availability, recovery time, recovery point objectives of each application?
- What is the architectural information of applications?
- What is the current service levels, change windows, and production issues?
- What is the unique security requirements for different applications?
- What is the total allocated UPS power, and % utilization?
- What is the battery backup time?
- What is the energy consumption patterns?
- What is the handling capacity of PDUs?
- what are the single points of failure?
- How much revenue will be lost by an unscheduled system outage

- What is the customer's tolerance for a service being not available?
- What is the number of racks of each category based on power ratings, cooling requirements, and size of hardware?
- What is the power and cooling requirement based on expected number of servers, network and storage devices to be hosted in the Data center?
- What is the type and capacity of network and security hardware to be deployed?
- What is the number of intranet/internet/VPN users accessing different shared services in the source Data center?
- What is the number of intranet/internet/VPN users accessing different shared services in the target Data center?
- What is the number of network hosts, mailboxes, backup clients, security zones needed in the target Data center?
- What is the network, server, storage, and database monitoring requirements in the target Data center?
- What are the assets (hardware/software) which will surpass the EOL?
- Estimation of total UPS and raw power requirement for the hardware design to be implemented at target data center?
- What is the number of racks & square footage needed in the target Data center?
- How many backup generators are there in the source Data center?
- Do you have any Data center energy usage monitoring system in the source Data center?

COBIT Checklist for Auditing Data Center

No	Audit Question	Status
1	Review data center exterior lighting, building orientation, signage, and neighborhood characteristics to identify facility related risks.	
2	Research the data center location for environmental hazards and to determine the distance to emergency services.	
3	Review exterior doors and walls to determine if they protect data centers facilities adequately.	
4	Evaluate physical authentication devices to determine if they are appropriate for the manner in which they are being used and are working properly.	
5	Review security guard building rounds logs and other documentation to evaluate the effectiveness of the security personnel function.	
6	Verify that sensitive areas are secured adequately.	
7	Verify that heating, ventilation, and air-conditioning systems maintain constant temperatures within the data center.	
8	Evaluate the data center's use of electronic shielding to verify that radio emissions do not affect computer systems or that system emissions cannot be used to gain unauthorized access to sensitive information.	
9	Determine whether the data center has redundant power feeds.	

10	Verify that ground to earth exists to protect computer systems.	
11	Ensure that power is conditioned to prevent data loss.	
12	Verify that battery backup systems are providing continuous power during momentary black-outs and brown-outs.	
13	Ensure that generators protect against prolonged power loss and are in good working condition.	
14	Ensure that a burglar alarm is protecting the data center from physical intrusion.	
15	Verify that a fire alarm is protecting the data center from the risk of fire.	
16	Ensure that a water alarm system is configured to detect water in high-risk areas of the data center.	
17	Ensure that a humidity alarm is configured to notify data center personnel of either high or low-humidity conditions.	
18	Review the alarm monitoring console(s) and alarm reports to verify that alarms are monitored continually by data center personnel.	
19	Ensure that data center building construction incorporates appropriate fire suppression features.	
20	Ensure that data center personnel are trained in hazardous materials handling and storage and that hazmat procedures are appropriate.	
21	Verify that fire extinguishers are placed every 50 ft within data center isles and are maintained properly.	
22	Ensure that fire suppression systems are protecting the data center from fire.	
23	Verify that surveillance systems are designed and operating properly.	

24	Ensure that physical access control procedures are comprehensive and being followed by security staff.	
25	Review facility monitoring procedures to ensure that alarm conditions are addressed promptly.	
26	Verify that network, operating system, and application monitoring procedures provide adequate information to identify potential problems.	
27	Ensure that roles and responsibilities of data center personnel are clearly defined.	
28	Verify that duties and job functions of data center personnel are segregated appropriately.	
29	Ensure that emergency response procedures address reasonably anticipated threats.	
30	Verify that data center facility-based systems and equipment are maintained properly.	
31	Ensure that data center personnel are trained properly to perform their job functions.	
32	Ensure that data center capacity is planned to avoid unnecessary outages.	
33	Verify that procedures are present to ensure secure storage and disposal of system media.	

Assessing The Applications in Data Center

Assessing the applications in data center involves establishing and tracking the relationships and dependencies between IT components such as servers, networks, storage, and applications that comprise an IT service.

Assessing the applications is an important activity in data center initiatives (like relocation, expansion, migration); which helps you avoid unforeseen incidents and problems in delivering IT services. It minimizes the risks associated with data center initiatives and provides the opportunity to optimize resources to reduce overall total cost of ownership (TCO).

It helps the data center migration planners in providing valuable data on application performance, dependent hardware, dependent business units, SLA's etc. enabling them to make informed decisions based on real data, not outdated architecture diagrams.

Assessing the resources, in Data center with respect to data applications, can be done by the below mentioned questions:

- What is the name of the application, app type, app environment and ID?
- What are the app interfaces?
- What is the risk associated?
- What is the current SLA, RTO, RPO and AO associated?

- What is the status of the app (Dev/Prod/QA)?
- What is the app database name, instance, hostname, vendor?
- Is the application hosted on an individual server? If yes, what is the hardware model, vendor, hostname, disk size, number of CPU's, OS type, OS version, RAM, and where is the server located (city and rack location)?
- How many virtual servers are involved, and what are their details?
- What is the IP address, and associated security zone?

Best Practices for Costs Reduction in DC

IT departments are under constant pressure from the top level management to reduce costs, and then it's left up to middle level management to determine how to meet those targets. IT cost reduction will become more challenging in DC environment; hence, here are some best practices for reducing costs in data centers:

- Eliminate the unnecessary assets (IT assets, software, maintenance and support, and disaster recovery contracts) from the Data center.
- Renegotiate all hardware, lease, software, maintenance and support contracts.
- Virtualization of hardware to reduce the costs in procuring assets and improve operational efficiency.
- Meticulous monitoring on software licenses and their expiry dates.
- Installation of high efficiency UPS can reduce the power and cooling costs.
- Implementation of IT asset management process and tool will enable to capture accurate information on the IT assets.
- Implementation of capacity & availability monitoring tools, to monitor the applications and the servers.
- Reduce the number of backup applications to meet recovery SLAs and reduce capital expenses, operating expenses, and training costs.

- Deploy de-duplication everywhere to help address the information explosion and reduce the rising costs associated with backing up data.
- Once in a quarter, do an audit to identify the redundant IT hardware assets, software, and contracts.
- Rationalization of the hardware to provide clear information on the assets that are useful and obsolete.
- Renegotiate all maintenance and support contracts to get rid of those that are costly.
- Data center consolidation is a popular approach in the last decade for achieving cost savings and other efficiencies.

Data Center Retrospective Review Report

DC Name:	**Location:**
DC Manager:	**Date of Review/Audit:**
Document Name: **Version :**	**Reviewed/Audited by:**
Project/Contract number	**Customer name:**

Reason for Retrospective Review:

Description of Methodology:
Documents Reviewed:
Individuals Interviewed:

Start Date :	End Date :
Description of the DC:	
Scope of Work:	

Team members:

Name	Role in ODC	Responsibilities

Reasons for DC Closure:

Best Practices Followed in the DC:
Customer Details: Customer Level of Cooperation: Customer Complaints: Penalization:

Challenges Faced: **Resolution Methods:**
Suggestions for Executing a Similar DC Engagement:

Review Findings:

Review Conclusions:

Data Center Maturity Level

What is a maturity model?

Maturity models is a business tool or methodology used to assess their current or as-is operations and processes and to identify the gaps in operations and processes (with respect to people, processes, and products) for effective operations in overall business.

Data Center Maturity Model

Data center maturity model is a business methodology which allows assessing the Data center processes and operations through some standard best practices from IT Service Management, IT Governance, and Project Management. Maturity is indicated by the award of a particular maturity levels like 0,1,2,3,4.

Maturity Level 0 - This generally denotes ad-hoc operations, no processes defined. Here firefighting happens after an incident, no monitoring tools, some tools exist but no defined process, cannot effectively identify data center assets, where the assets are located, or how devices are connected without significant manual intervention.

Maturity Level 1 - This generally denotes reactive operations. Here there are some informal processes but no appropriate documentation. Data center operations rely on the heroics of some individuals, and monitor some key components of the data center.

Maturity Level 2 - This generally denotes defined and proactive operations. Here there is defined ITSM and ITAM processes like incident, problem, change, configuration, security, inventory, disposal management processes. Also there will be monitoring tools that track resources, but do not provide very detailed information on space, power, cooling, port connectivity, equipment, etc. These organizations lack a service-driven approach where data center activities should be aligned to business goals and objectives.

Maturity Level 3 - This generally denotes stable and managed operations. Here the operations are well managed ensuring the services are delivered with respect to defined SLA's and OLA's, processes are robust and integrated across the complete service lifecycle.

Maturity Level 4 - This generally denotes optimized operations. Here the Data center operations is having a holistic view of the data center, issues are identified before they occur, implementation of automation, and running operations as per the Green IT practices considering the environment and efficiency in managing the assets with the focus on cost reduction.

This assessment is carried out on different categories of DC operations and facilities like computing, storage, network, power, cooling, etc.

Green Data Center

Data centers that use energy-efficient servers, networking devices, mechanical equipment, lighting equipment, electrical equipment targeting reduction in energy demands.

The primary objectives of green data centers are:

- To maximize energy efficiency
- Reduces energy costs
- Reduce carbon emissions and environmental impact

Best Practices for Green Data Center

- Block cable openings to prevent cold air waste in the hot aisle.
- Removing under-floor cable blockages that impede airflow.
- Turning off servers that are not doing any work.
- Turning off computer room air conditioning (CRAC) units in areas that are overprovisioned for cooling.
- Organizing IT equipment into a hot aisle and cold aisle configuration.
- Positioning the equipment so that you can control the airflow between the hot and cold aisles and prevent hot air from re-circulating back to the IT equipment cooling intakes.
- Implementation of MAID technology.
- Procuring smart energy assets, considering humidity, temperature and compliance with green regulations.

- Installation of catalytic converters on backup generators.
- Use of alternative energy technologies such as photovoltaics, heat pumps, evaporative cooling, low-power lamps, and lighting sectors.
- Use of low-emission building materials, carpets, and paints.
- Server and rack management for better air-flow; for optimizing the airflow, with intelligent fans close to IT source.
- Use of Sea Water/Recycled Water and reuse of heat into office buildings where applicable.
- Cooling the IT equipment within the racks.
- Reduction of disk drives and server hardware which will significantly lower energy consumption in file serving environments.
- Using natural ways to cool equipment.
- Collection of data manually or through monitoring systems to determine the state of your current systems. Determine how much power is consumed by devices and applications, and break down the costs.
- Energy usage should be continuously monitored to determine peak and low energy demands.
- The energy savings plan should be documented and rewarded. Also, it should be reviewed regularly with corrective action taken to address failures.
- An inventory of your current systems, and tracking their power usage and locations.
- Track your company's business and growth plans - to help forecast future needs.
- Establish goals for reducing your company's carbon footprint - and the timeframe set for achieving those goals.

Advice and Interview Questions for Data Center Management Roles

Interview Questions for Data Center Management

What are the general equipment that you know in data center?

General equipment found in data center are servers, storage equipment, network equipment, racks, network cables, uninterruptible power supply, transformers, power distribution units, power cables, backup generators, chillers, computer room air conditioning units, direct expansion air handler units, pumps, cooling towers, etc.

What is the difference between Tier 3 and Tier 4 Data centers?

A Tier III data center offers 99.98% availability and Tier 4 data center can provide, with 99.99% availability. Tier III data centers do not have total modular redundancy.

Tier IV Data center architecture can withstand even the most serious of technical incidents without server availability ever being affected.

What happens when there is no cable management in Data center?

Lack of cable management in Data center leads to:

- Unmanaged and unstructured connections leading to confusion
- Heat sink creation (blocking of air flow)
- Difficulty in identifying which cable is meant for which applications
- Dead ports in switches and cable leads to unused capacity
- Consumption of extra floor space

What are the key considerations for cables in Data center?

- Cable span (within racks or within and between racks)
- Cooling and airflow
- Power consumption
- EMI (Electromagnetic interference)
- Cost
- High quality, reliability, and scalability
- Support for storage devices (NAS, SCSI, etc.)
- Support for higher speed technologies.

Why is it not good to outsource Data center?

For big conglomerates, it is advisable to have its own Data centers since:

- Its server environment contains company's most valuable items and handles business critical functions, so it is advisable to be taken care by its employees.
- No one can know your company's server environment needs like your own people who are dedicated to supporting it.
- No matter how good an outside vendor is, it does not have a personal stake in making sure that your Data center runs correctly the way that you and your coworkers do.

What are the best conditions to outsource a Data center?

It is good to outsource when:

- You have a short-term need for a server environment, perhaps until a permanent Data center is constructed.
- You want a standby facility ready to take over for a primary Data center in the event of a catastrophic event.

What are the advantages of blade server?

Advantages of blade servers are:

- It has a higher calculation capacity, higher stability, scalability, versatility, performance, availability, easy serviceability, simple systems management and deployment, and it occupies less space than other servers.

- Reduces the cost and the working temperature.
- Network access for all server blades in a chassis reduces cabling complexity.

What are the advantages of tower server?

Advantages of tower servers are:

- Easier cooling.
- Scalability to add number of servers to the existing network.
- Ideal for small and medium sized business.

What are the advantages of rack server?

Advantages of rack server are:

- Easy to add servers to the rack, also offers an adjustable mounting depth, allowing the rack to be used with multiple server sizes.
- As rack servers are enabled with lockable sides and doors, the rack helps improve security.
- Enables proper airflow, where you avoid multiple problems, such as overheating, improving the performance of your servers.
- Ability to work with a wide variety of servers, and it provides an ability to be compatible with equipment from multiple vendors.
- Maintenance is easy as the shelves ensure that servers are easily accessible, making routine maintenance a breeze.

What are the different migration methods that happen in Data centers?

- Tape migration
- IP based Host based migration using replication
- Host based migration
- Host based migration using server virtualization
- Host based migration using file copy and NAS
- Appliance based migration

What would be some of the criteria for selecting servers?

Selection criteria:

- Should not be too tall that access is difficult or gets too close to the roof. Heat rises and equipment at the top will be subject to higher temperatures
- Computer hardware standard is 19 inches and networking hardware NEBS standard is 21 inches (Network Equipment Building Standards).
- Must be deep enough for your equipment to fit plus enough space for vertical and horizontal cabling.
- Cables and equipment should not protrude into aisle space.
- Should have in-built fans.

What are the benefits of automation in Data centers operations?

- Reduced dependency on IT staff
- Improved IT-process quality

- Simplified regulatory compliance

How much does it cost for an outsourced Data center?

Costs for an outsourced Data Center are usually dictated by the amount of floor and rack space your servers occupy, how much power they draw, and what level of connectivity and staff support they require.

What can be the 4 essential design strategies for well-designed Data center architecture?

It can be mentioned as:

- It should be robust – standby power supply, redundancy in equipment, no single point of failures, etc.
- It should be modular – sufficient power, cooling through the rooms.
- It should be flexible – build infrastructure systems using components that are easily changed or moved.
- It should be standardized – using reliable technology, vendors, IT best practices, and standards.

What are the different costs that occur in a DC outage?

Recovery costs: Costs that relate to bringing the organization's networks and core systems back to a state of readiness.

Equipment costs: Cost that are related to new equipment purchases, repairs, including refurbishment.

Third-party costs: The cost of other 3rd party company staff members/specialists engaged to resolve the unplanned outages.

Loss of revenue: The total revenue loss from customers and potential customers because of the inability to adhere to SLA's.

Penalization: The total amount paid to customers and potential customers because of the inability to adhere to SLA's.

IT productivity loss: The lost time and related expenses associated with IT personnel downtime.

Employee overtime costs: The costs incurred due to employee overtime.

What are the primary capacity planning components for Data center?

- Rack optimization
- HVAC
- Power utilities
- Servers and virtual machines

What can be the different causes of a DC outage?

- IT equipment failures
- UPS system failures
- CRAC failures
- Generator failure, Weather-related
- Accidental/human errors

Difference between SMF and MMF

Single Mode cable is a single stand of glass fiber with a diameter of 8.3 to 10 microns that has one mode of transmission. Single mode carries higher bandwidth than multimode fiber, but requires a light source with a narrow spectral width. Single-mode fiber gives you a higher transmission rate and up to 50 times more distance than multimode, but it also costs more. Single-mode fiber has a small light carrying core of 8 to 10 microns in diameter. It is usually used for long distance transmissions.

Multimode cable is made of glass fibers, with common diameters in the 50-to-100 micron range for the light carry component. Multimode fiber gives you high bandwidth at high speeds over medium distances. Multimode fiber has a relatively large light carrying core, usually 62.5 microns or larger in diameter. It is usually used for short distance transmissions with LED based fiber optic equipment.

When is it good to use SMF and MMF?

If the distance is less than a couple of miles, multimode fiber will work well. If the distance to be covered is more than 3-5 miles, single mode fiber is the choice.

What is the difference between Data center and NOC?

Data center is the center where all the data is stored through storage, networking, and high end computing. NOC is a center where all the administrators supervise and monitor the functioning of the IT systems.

Why should there be an inventory for Data center?

- To create a baseline with set of accurate information for all assets.
- To locate, control, and account the assets.

As an auditor what are the things that you will focus on while auditing?

- Overall Data Center Design Strategy
- UPS Design Strategy
- HVAC Design Strategy
- Fire Suppression Design Strategy
- Standby Generator Design Strategy
- Building Management System (BMS) Design Strategy

- Building Architecture Design Strategy
- Facility Staffing Practices
- Data Center Facility Technology Training Methodologies
- Vendor Management Program Practices
- HVAC Maintenance Practices
- Electrical Maintenance Practices
- UPS Maintenance Practices
- UPS Battery Maintenance Practices
- BMS Maintenance Practices
- Standby Generator Maintenance Practices
- Fuel System Maintenance Practices
- Fire System Maintenance Practices
- Data Center Facility Definition of Services
- Facility Management Problem Management Process
- Facility Management Strategy
- HVAC System Management Strategy
- Electrical System Management Strategy
- Data Center Facility Technology Tool Usage

When the servers are overheating, and what are the possible measures you can take?

- Plan your setup and cooling fan locations to achieve the best airflow possible.
- Eliminate clutter in the room to optimize airflow and circulation. This also makes the room safer and easier to work in when you need to reach those hard-to-get-to areas.
- Try to keep the temperature in the room below 77°F. The cooler it is, the better.

- If there are empty spaces on your server rack, install blanking panels to fill them. A rack with empty spaces will reduce airflow.
- Consider utilizing a rack enclosure that has a built-in cooling system.
- Employ a solution to monitor the temperature in the server room. Some temperature monitors will even email you when the room gets too hot.
- Use a redundant power supply designed for your specific equipment.

What are the key attributes of distributed Data center?

- Load balancing across sites and high availability – which leads to increased availability and performance.
- Optimal routing and response times – which leads to better response times and customer satisfaction.

What is the difference between Data center and cloud?

Cloud is an off-premise form of storing data on an off-premise computer/computers through internet. And a data center refers to on premise hardware that stores data within an organization's local network.

When would you need a Data center and a cloud?

Data center is ideal for companies where there is sensitive data in the information systems and needs customized, dedicated system that gives them full control over their data and equipment.

Cloud system is good for companies where the data is not secretive and sensitive, and where there are budget constraints and whose requirements vary time to time.

What are the standards defined for Data center infrastructure?

TIA 942 -Telecommunications Industry Association (TIA) is the standard that defines guidelines for planning and building data center, including cabling, facility, network design, and data center tiers.

TIA 607 - Telecommunications Industry Association (TIA) is the standard defined for planning, installation, and design of Data center telecommunications system.

Why is virtualization used in cloud computing?

Virtualization is the main cost-saving, hardware-reducing, and energy-saving techniques used in cloud computing.

Virtualization is implemented through software that will enable the underlying physical machine into different virtual machines (VMs).

What are the benefits of cloud computing in comparison with Data center?

Cloud computing provides benefits like:

- Flexibility – There is the ability to update hardware and software quickly to adhere to customer demands and updates in technology.
- Savings – There is a reduction of capital expenditures and IT personnel.
- Location & Hardware Independence – Users can access application from a web browser connected anywhere on the internet.
- Multi-tenancy – Resources and cost are shared among many users, allowing overall cost reduction.
- Reliability – Many cloud providers replicate their server environments in multiply data centers around the globe, which accounts for business continuity and disaster recovery.
- Scalability – Multiply resources load balance peak load capacity and utilization across multiply hardware platforms in different locations
- Security – Centralization of sensitive data improves security by removing data from the users' computers.
- Cloud providers also have the staff resources to maintain all the latest security features to help protect data.
- Maintenance – Centralized applications are much easier to maintain than their distributed counter parts.

All updates and changes are made in one centralized server instead of on each user's computer.

Can you tell some key considerations that you will take before Data center migration?

- Relocating DR instance of all mission critical applications to another data center in a different seismic zone.
- Ensuring that availability and recoverability of the applications is not impacted in transition.
- Ensuring that all interdependent applications are able to communicate with each other in view of mission critical applications being relocated while other being retained.
- Adopting virtualization & consolidation technologies.
- Ensuring new facilities infrastructure is capable of providing power and cooling to high density IT infrastructure.
- Ensuring optimal costs are incurred on hardware software and labor.

What is the difference between availability and recoverability?

The difference is the ability of a CI or service to perform its required task at a stated period of time is availability.

The process of restoring business services after disruption of service is by transferring IT resources to secondary or backup locations.

What is the difference between a server and a desktop?

- Servers are machines which provides services to support multiple users. They are reliable machines designed to support multiple users and run a large number of different services and applications. They are meant for managing, storing, sending, and processing data 24-hours a day. Servers generally run more powerful operating systems that can handle networking, email, internet/intranet hosting, file sharing, databases, and more. Server machines are equipped with better hardware configurations, redundancy. Servers impact huge number of people.
- Desktops are simple machines which have user friendly operating systems and applications to facilitate simple activities. They are single or dual processor machines, and have single hard drive and power supply.
- Servers have multiple processors, hard disk drives, RAM slots, SMPS.
- Desktops have single or dual processor socket, 1 hard disk drive, 2-4 RAM slots.
- Servers can work without shutting down for years.
- Desktop has to be shutdown at regular intervals.
- In servers, if 1 SMPS becomes hot another SMPS will be turned on and the heated SMPS will be turned off automatically.
- In desktops, if 1 SMPS is hot it has to be turned off.
- In servers, if 1 HDD is crashed the systems are still available
- In desktops, if 1 HDD is crashed - you cannot access the machines. You will have to reinstall the operating system.

How do you calculate the cost of an outsourced Data center?

There are no magical calculators which can estimate the costs for an outsourced Data center, but it is usually approximated by the amount of floor and rack space your servers occupy, how much power they draw, and what level of connectivity and staff support they require.

How do you think virtualization benefits Data center operations?

Virtualization helps in numerous ways and some of the important ones are:

- Lessens the heat produced
- Reduces the amount of hardware costs
- Redeployment time of a server is quick, in minutes
- Backup of the servers is quick and easy
- Reduces the carbon footprint
- Disaster recovery is easier and with much success rate

What are the different network cabling architectures available?

Centralized: Switches are consolidated in a centralized MDA and HAD row.

Top of Rack: Switches are located within the EDA cabinet or rack.

End of Row: Switches are located at the end of the network row within an HAD cabinet.

Middle of Row: Switches are located in the middle of the network row within an HAD cabinet.

What are the pros and cons of centralized cabling architecture?

Pros:

- Simple in design, easy for maintenance
- Good port utilization

Cons:

- Too many cables & cable overlaps
- Lack of scalability

What are the pros and cons of EOR/MOR cabling architecture?

Pros:

Fewer number of cables

Good scalability

Cons:

Increased management overhead

Network stability risks

What are the pros and cons of TOR cabling architecture?

Pros:

Efficient use of floor space

Good scalability and easy maintenance

Cons:

More switches are needed

Higher equipment costs

If you are planning to buy used servers, what are the parameters that you have to focus upon?

Important parameters would be:

- Is the equipment upgradable?
- Does the equipment support multi-processor?
- Does the equipment support RAID?

What are the generic server maintenance tasks?

Important maintenance tasks would be:

- Verifying backups are working
- Verifying disk usage
- Checking RAID alarms
- Checking server utilization
- Checking hardware errors
- Checking application updates and OS updates
- Review user accounts

What are the policies that are must for a Data center?

Some basic policies that should be maintained for Data centers are:

- Access Control Policy
- Escorted Access Policy
- General Cleanliness Policy
- Change management policy
- Asset retirement policy
- Safety policy
- Capacity planning policy

What are the generic attributes that you would look in servers?

Generic attributes can be mentioned as:

- Processor speed
- Chipset
- Memory
- Hard drives
- Network controllers

- Expansion cards
- RAID
- Power supply
- Dimensions

Why should a company invest in green Data center?

- To maximize energy efficiency
- To reduce energy costs
- Less use of resources
- To reduce carbon emissions and environmental impact

What is the key criterion for evaluating Data center?

- Data center services
- Data center infrastructure
- Redundancies
- Connectivity
- Uptime

What are the common issues faced by Data center managers?

- Heat density
- Space constraints

- Power density
- Availability
- Technology changes
- Energy cost and Energy efficiency

What are the typical risks in Data center migration?

- Unexpected or extended downtime
- Budget overruns
- Knowledge transfer issues

What are the forces of change for Data center projects?

- New safety and security regulations
- New power and cooling requirements
- New technology requirements like redundancy, etc.
- New/changed operational procedures
- Changes in floor spaces

What are the top facility issues that can be found in Data center?

- Excessive heat
- Insufficient power
- Excessive facility cost
- Insufficient raised floor area

What are the examples of capital and operating expenses that occur in Data centers?

Capital expenses: Server hardware, Network hardware, Storage hardware

Operating expenses: Installation costs, Server maintenance and network costs, Power costs, Software costs, Labor costs, etc.

What are the parameters for choosing the rack size?

- Height (Common standard heights for floor-standing racks and rack enclosures are 42U, 45U, 48U, and up to 58U)
- Width (Common standard width for racks are 24 inches or 600 mm and 30 inches or 750 mm)
- Depth (Common standard depth for racks are 42 inches and 48 inches, shallow racks 36 inches and 32 inches, wall mount racks 12, 16, 20, 32 inches)
- Load rating

What are the problems created by poor cable management?

- Signal interference and crosstalk due to improper placement of data and power cables
- Lengthy power and data cables

- Cable damages due to improper bends, movement, or sharp edges
- Cooling and airflow restriction
- Increased time for maintenance and troubleshooting the cables

How does a rack enclosure keep the equipment cool?

- side panels prevent hot air from re-circulating around the sides of the enclosure
- blanking panels force cold air through your equipment and prevent hot air from re-circulating through open spaces.
- thermal ducts ensure hot air is physically isolated, so there's no way for it to re-circulate and pollute the cold air supply.
- active heat removal accessories assist passive heat removal with ventilation fans

What are the parameters which will say that a data center is efficient?

- Carbon emissions impact
- Energy cost
- Operational costs

What are the monitoring measures that you will take for effective and efficient data center operations?

- Checking the temperature in data center
- Monitoring power and rack conditions
- Detecting fluid leaks
- Monitoring energy efficiency

What are the basic things that a data center provider should offer?

- Robust & scalable Data center design
- Fail safe security
- Redundant infrastructure
- Reliable power monitoring options
- Disaster recovery capability

Appendix

Acronyms

AC – Alternating Current

ACE – Availability, Capacity, & Efficiency

AO – Availability Objectives

ASHRAE – American Society of Heating, Refrigerating and Air-conditioning Engineers

ASP – Application Service Provider

ATS – Automatic Transfer Switch

BIA – Business Impact Analysis

CDN – Content Delivery Network

CFA – Core Facility Area

CFD – Computational Fluid Dynamics

CMDF – Containerized/ Modular Data center Facility

CSF – Critical Success Factors

CSW – Core Switch

CRAH – Computer Room Air Handler

CRAC – Computer Room Air Conditioner

DAS – Direct Attached Storage

DCC – Data center Consolidation

DCBX – Data Center Bridging Extension

DCIM – Data center infrastructure management

DCSP – Data center Service Provider

DCM – Data center Migration

DCN – Data Center Network

DCT – Data center Transformation

DBMS – Database Management System

DC – Direct Current

DCBX – Data center Bridge Exchange

DRP – Disaster Recovery Plan

EDA – Equipment Distribution Area

EEDC – Energy Efficient Data Center

EoR – End of Row

EPO – Emergency Power Off

FC – Fiber Channel

FCOE – Fiber Channel Over Ethernet

FSW – Fabric switch

GBIC – Gigabit Interface Converter

HA – High Availability

HDA – Horizontal Distribution Area

IDS – Intrusion Detection System

IPC – Inter Process Communication

ISO – International Organization for Standards

ITSM – Information Technology Service Management

KB – Knowledge Base

KPI – Key Performance Indicators

LAN/VLAN – Local Area Network/ Virtual Local Area Network

LEED – Leadership in Energy and Environment Design

MAID – Massive Array of Idle Disks

MDA – Main Distribution Area

MDS – Multilayer Director Switch

MEP – Mechanical, Electrical and Plumbing

MMF – Multi Mode Fiber

MSA – Master Software Agreement

NAS – Network Attached Storage

NEBS – Network Equipment Building System

ODC – Offshore Development Center

OLA – Operational Level Agreement

P2P – Physical to Physical

P2V – Physical to Virtual

PDU – Power Distribution Units

RACI – Roles, Accountability, Consulted, and Informed

RAF – Raised Access Floor

RCA – Root Cause Analysis

RCI – Rack Cooling Index

RPO – Recovery Point Objectives

RSW – Rack switch

RTO – Recovery Time Objectives

SAN – Storage Area Network

SATA – Serial Advanced Technology Attachment

SLA – Service Level Agreement

SLT – Service Level Targets

SMF – Single Mode Fibre

SOP – Standard Operating Procedures

SPOC – Single Point of Contact

TGG – The Green Grid

TIA – Telecommunications Industry Association

ToR – Top of Rack

UC – Underpinning Contracts

VBF – Vital Business Function

ZDA – Zone Distribution Area

Index

Made in the USA
Monee, IL
04 November 2026

67385826R10125